Teaching Mathematics to English Language Learners

Luciana C. de Oliveira · Marta Civil
Editors

Teaching Mathematics to English Language Learners

Preparing Pre-service and In-service Teachers

palgrave
macmillan

Editors
Luciana C. de Oliveira
School of Education
Virginia Commonwealth University
Richmond, VA, USA

Marta Civil
Department of Mathematics
University of Arizona
Tucson, AZ, USA

ISBN 978-3-030-48354-8 ISBN 978-3-030-48355-5 (eBook)
https://doi.org/10.1007/978-3-030-48355-5

Cover illustration: Witthaya Prasongsin

This Palgrave Macmillan imprint is published by the registered company Springer Nature Switzerland AG
The registered company address is: Gewerbestrasse 11, 6330 Cham, Switzerland

Contents

Notes on Contributors

Rachel S. G. Bower, Ph.D. is the Assistant Professor of Mathematics Education at Nevada State College. She holds a doctorate degree in mathematics education from Texas State University. She prepares hundreds of pre-service elementary school students each year in mathematics content and methods coursework. Her research interests include students and teachers in multilingual, secondary mathematics classrooms. She is a member of TODOS: Mathematics for ALL and NVTESOL.

Marta Civil is a Professor of Mathematics Education and the Roy F. Graesser Chair in the Department of Mathematics at The University of Arizona. Her research looks at cultural, social, and language aspects in the teaching and learning of mathematics; participation in the mathematics classroom; connections between in-school and out-of-school mathematics; and parental engagement in mathematics. Her work is situated in working-class, Mexican American communities and is based on a funds of knowledge orientation. She has led several funded projects working with children, parents, and teachers, with a focus on developing culturally responsive learning environments in mathematics education.

Gerardo De La Garza earned a B.A. in History with an emphasis in secondary education from the University of Northern Colorado, an M.A. in Curriculum and Instruction with an emphasis in English as a Second Language from the University of Colorado Denver, and an M.A. in Educational Leadership from the University of Colorado Denver. Mr. De La Garza has dedicated twenty-five years to the field of education as a teacher, dean, assistant principal, and principal, serving in Denver Public Schools and Aurora Public Schools. Mr. De La Garza is the principal of the school described in the chapter during the turnaround phase.

Luciana C. de Oliveira, Ph.D. is Associate Dean for Academic Affairs and Professor in the School of Education at Virginia Commonwealth University. Her research focuses on the role of language in learning the content areas and teacher education, advocacy and social justice. Dr. de Oliveira was the first Latina to ever serve as President (2018–2019) of TESOL International Association.

Melinda Eichhorn, Ed.D. is an assistant professor at Gordon College and a mathematics specialist in the Learning Disabilities Program at Boston Children's Hospital. Mindy currently teaches courses required for special education licensure (Pre-K to Grade 12) and supervises undergraduate teacher candidates.

Melissa A. Gallagher is an assistant professor of Mathematics Education at the University of Louisiana at Lafayette. Her research focuses on supporting teachers to develop the knowledge, beliefs, and practices necessary to support the mathematical potential of all students, especially English language learners.

Karen M. Gregory is an assistant professor and Director of TESOL at Clarkson University, Capital Region Campus. Previously, Dr. Gregory taught ENL to newcomer high school students, co-taught ENL science and social studies, and also taught high school Spanish. Dr. Gregory's research is focused on best practices of teaching English as an additional language, TESOL professional development, and social-ecological approaches to whole school improvement. Dr. Gregory has published in the *TESOL Encyclopedia of English Language Learning*, *NYS TESOL Journal*, *The Journal of School Leadership*, and the *Journal for Leadership and Instruction*, and has contributed to several book chapters in her field.

Michelle Hale is a classroom teacher in Valley View School District. She is a fourth-year, licensed, secondary mathematics classroom teacher with a B.A. in mathematics. Michelle continues to contribute to the mathematics education community by providing professional development in national mathematics conferences over the past four years. Her research has primarily consisted of research-based pedagogical practices in mathematics education. She is interested in exploring how to deliver rigorous mathematical content to students of highly diverse strengths and needs.

Holly Hansen-Thomas, Ph.D. is Professor of ESL and Bilingual Education and Associate Dean for Research and Scholarship in the College of Professional Education at Texas Woman's University in Denton, TX. A two-time Fulbright Scholar and successful grant writer/manager of approximately 7 million dollars in federal funding, Dr. Hansen-Thomas has engaged in teacher training and research for content area teachers of Emergent Bilinguals (EBs) in the United States, Europe, and Central America, for nearly two decades. Her research interests include ESL training for mainstream secondary level teachers; EBs' development of academic language in mathematics and science; language awareness; and teacher identity.

Susan Holloway has earned the degrees of B.A.E. from the University of Arizona, a B.S. in Engineering from the University of Tennessee, an M.A.E degree from CU Boulder, an M.S. in Mathematics from the University of Texas, and an Ed.D. from CU Denver. Her academic and professional practice focuses on formulating successful educational leadership for ELLs in the mathematics classroom. Presently, Dr. Holloway is a high school mathematics instructor and department head, as well as an adjunct mathematics instructor at CU Denver and at the Aurora Community College. She serves as the Political Liaison for the Colorado TESOL (CoTESOL) organization.

Ji Yeong I, Ph.D. is an assistant professor in School of Education at Iowa State University. She teaches secondary mathematics methods courses focusing on Algebra, Problem-Solving, and teaching mathematics for Emergent Bilinguals. Her research agenda embraces the search for an effective mathematical pedagogy for culturally and linguistically diverse learners in the United States and South Korea. Her current project focuses on implementing mathematical modeling to support Emergent Bilinguals to understand rigorous mathematics.

Juliet Langman, Ph.D. is Professor of Applied Linguistics in the Department of Bicultural-Bilingual Studies and Senior Associate Dean of Graduate Studies and Operations in the College of Education and Human Development at UTSA. Her research interests focus on minority youth populations in multilingual settings, exploring the intersection between language use, learning, and identity. In her work, she draws on data from neighborhood, community, and school-based contexts in Central Europe and South Texas. Her current school-based work focuses on the needs of English language learners and their teachers in STEM content area classrooms.

Lawrence M. Lesser is a UTEP Distinguished Teaching Professor, a former high school teacher, and a founding editor of *Teaching for Excellence and Equity in Mathematics*. In *Statistics Education Research Journal*, his research team published the first major research on ELLs in statistics.

Amanda E. Lowry holds a Ph.D. from Rutgers University and has been teaching ESL math at a New Jersey high school for over 8 years. She specializes in ways to promote mathematics learning for recently arrived adolescents by focusing on the educational experiences they bring to their new school while bridging the transition to a new educational setting and culture.

Vanessa Z. Mari, Ph.D. is the Assistant Professor of TESL at Nevada State College. She holds a doctorate degree in bicultural and bilingual studies with a concentration in culture, literacy, and language from the University of Texas San Antonio. She has also worked as the English Language Fellow in Piura and Lima, Peru and was a Fulbright scholar in Uruguay. Her research interests include translanguaging, teacher preparation, and best practices for English language teaching.

Carla Meskill is Professor, Department of Educational Theory and Practice, State University of New York, Albany. Since the first desktop computers emerged in the 1980s, she has been at the forefront of design, research, and practical training in how digital screens can best be made use of in education. Her publications include numerous books and research articles on digital screen mediation in classrooms, in online instruction, and in educator professional development. Her most current book, *Digital Screen Mediation: Agency and Authenticity with Instructional Technologies*, is available in December 2020.

Alandeom W. Oliveira is an associate professor of science education at the State University of New York at Albany. He earned a Master's degree in science education at Southeast Missouri State University (2002) and a Ph.D. degree in science education at Indiana University Bloomington (2008). He has taught science education courses to teachers in Brazil and the United States and has coordinated multiple professional development programs for school teachers, including Science Modeling for Inquiring Teachers Network, and Technology-Enhanced Multimodal Instruction in Science and Math for English Language Learners. His research interests include cooperative science learning, inquiry-based teaching, and classroom discourse and language use.

Gretchen P. Oliver, Ph.D. is an assistant professor and Assistant Director of TESOL Programs at Clarkson University, Capital Region Campus, where she teaches TESOL courses and supervises pre-service teachers. She was the lead field researcher and project coordinator for a US Department of Education research and professional development grant-funded project. Additionally, she was a K-12 French teacher and taught EFL in France. Her research interests include teacher education and professional development, content-based language instruction, and culturally responsive teaching. Dr. Oliver has published in the *TESOL Encyclopedia of English Language Learning* and has contributed to several book chapters in her field.

Alexander Radosavljevic, Ph.D. is a Course Instructor at Western Governors University (WGU) for Mathematics for Elementary Educators courses. He has been a high school teacher of mathematics, English, and Special Education in the Chicago Public Schools for 14 years. Alex has also been an evaluator of K-12 school programs, director of an outdoor education STEM program, and has taught mathematics content and teaching methods courses at several universities. His research interests include the intersection of language, culture, and STEM education, assessment, and out-of-school learning.

Sarah A. Roberts, Ph.D. is an assistant professor of mathematics education at University of California, Santa Barbara. Her research focuses on issues of equity in mathematics education, specifically with regard to multilingual students. Dr. Roberts is interested in preparing effective, equity-focused pre-service and in-service teachers who support students; and she studies those teachers' learning experiences.

Maria I. Ruiz is an assistant professor of Special Education at the University of Louisiana at Lafayette. Her research focuses on developing responsive teachers who can effectively address the needs of diverse students, including students with disabilities and English language learners, and their families.

Angela R. Thompson, Ph.D. is an Associate Professor of mathematics education at Governors State University. In addition to teaching undergraduate mathematics courses, she teaches content pedagogy and prepares pre-service teachers. This includes both secondary mathematics pre-service teachers as well as K-12 candidates in other content areas. Her research interests explore what happens at the intersection of language, culture, and mathematics in a classroom, and activism in social justice and equity in K-16 mathematics education.

Sepideh Yasrebi is a Ph.D. candidate at State University of New York, Albany. She received her M.S. Ed. from Hofstra University in New York in 2015. She investigates narrative competence and cognitive mapping of English language learners in a US K-12 school. She was an ENL/EFL teacher in Long island, Bronx, Queens, and Iran. Her research interests lie in discourse analysis with social theory, critical curriculum theory, critical race theory, intercultural pragmatics, and semantics-pragmatics interface.

Jiaqi Yu is a Ph.D. candidate at Iowa State University majoring in Instructional Technology and co-majoring in Human-Computer Interaction. Her research has been residing at the intersection of online learning, instructional design, and emerging technology in teaching and learning. Some of her recent studies have been focusing on preparing K-6 teachers for technology integration and helping pre-service teachers broaden their understanding around makerspaces and maker pedagogy.

List of Figures

List of Tables

1

Introduction

Luciana C. de Oliveira and Marta Civil

Students identified as English language learners (ELLs) are a rapidly growing population in the United States, accounting for 9.6% or 4.9 million students in public K-12 schools (National Center for Educational Statistics [NCES], 2019). The largest number of students classified as ELLs are found in California (20.2%), Texas (17.2%), Nevada (15.9%), New Mexico (13.4%), and Colorado (11.7%) (NCES, 2019).

With such growth in the ELL population, teachers of all content areas are seeing more of this student population in their classrooms. Preparing all teachers for the learning needs of ELLs—not just ESL teachers or bilingual specialists—is a pressing concern (Lucas & Grinberg, 2008). Teacher education programs across grade levels and content areas need to adequately prepare teacher candidates with knowledge of

L. C. de Oliveira (✉)
School of Education, Virginia Commonwealth University, Richmond, VA, USA

M. Civil
Department of Mathematics, The University of Arizona, Tucson, AZ, USA
e-mail: civil@math.arizona.edu

L. C. de Oliveira and M. Civil (eds.), *Teaching Mathematics to English Language Learners*, https://doi.org/10.1007/978-3-030-48355-5_1

specific methods, curriculum, instruction, and assessment for the needs of ELLs (de Oliveira, 2016). Content teachers, for example, are expected to be able to provide disciplinary-based instruction that goes beyond a focus on vocabulary (Moschkovich, 2013; Schleppegrell, 2004). This is particularly evident in the recent development and adoption of the Common Core State Standards for English Language Arts and Literacy in History/Social Studies, Science, and Technical Subjects (CCSS-ELA/Literacy), the Common Core State Standards for Mathematics, the C3 Framework for Social Studies State Standards, and the Next Generation Science Standards (NGSS). These new standards highlight the role that teachers have in developing ELLs' discipline-specific abilities.

In mathematics, the Common Core State Standards (National Governors Association Center for Best Practices & Council of Chief State School Officers, 2010) movement has increased the linguistic demands associated with learning mathematics, focusing on such high-level skills as engaging in meaningful discussions, expressing ideas beyond computation skills, reading and writing complex word problems, and using language to express mathematical understanding. This increased focus on higher-order language and literacy tasks intensifies the needs of ELLs in the mathematics classroom.

As the chapters in this book attest, while mathematics teacher educators recognize the importance of educating pre-service and in-service teachers to teach ELLs, guidelines for how to do so remain scarce in the literature. What does mathematics pre-service teacher preparation look like when it meaningfully incorporates preparation to teach ELLs? How do mathematics curricula change when teacher educators address ELL learning? How can mathematics in-service teachers learn to work more effectively with ELLs in their current classrooms?

Terminology and Acronyms

Putting together an edited volume like this requires authors to consider the terminology used to identify the target student population with whom they set out to work. We recognize that there is little agreement in the scholarly literature as to what name best describes this student population. Each designation has different connotations and issues, with different terms being favored by scholars. These include emergent bilingual, bi/multilingual students, additional language speaker, ELL, English learner (EL), Limited English Proficient (LEP), non-native speaker (NNS), L2 speaker, etc.

We left it open for authors to choose whatever terminology was most appropriate for their chapters, without imposing a particular term or acronym. We use "English Language Learners" (ELL) in the book title to achieve cohesiveness among other books in this Palgrave Macmillan mini-series (see de Oliveira & Obenchain, 2018; de Oliveira & Shoffner, 2016; de Oliveira & Wilcox, 2017).

Organization of the Book

Chapter 2, by Ji Yeong I and Jiaqi Yu, opens the book with a case study that investigated how middle school mathematics pre-service teachers (PSTs) help Emergent Bilinguals (EBs) understand cognitively demanding problems while they received a set of EB-focused interventions. This chapter, entitled "Preparing Mathematics Preservice Teachers for Teaching Emergent Bilinguals Through Concurrent Intervention", examines how three PSTs, who had no prior experience of teaching mathematics to EBs, worked with EBs in a one-on-one setting while receiving individual interventions. As time evolved, the PSTs began to integrate content from students' cultures and applied various alternative methods related to the given mathematical situations. This chapter suggests that preparing PSTs requires an infusion of practical experiences and examples in order to teach EBs with effective culturally responsive teaching strategies.

In Chapter 3, Lawrence M. Lesser reports on his efforts to redesign a university's statistics literacy course to give its biggest clientele (preservice elementary and middle school teachers) content and pedagogy to improve their learning and make their future teaching more effective in a geographic region with a significant proportion of ELL students. The research-informed redesign made the course more accessible and useful and yielded insights about the distinctive ways in which ELL-responsive pedagogy plays out in the statistics classroom, a context that had received little specific attention (compared to the mathematics classroom) despite its rapidly growing importance in curricular and societal trends.

Chapter 4, by Maria I. Ruiz and Melissa A. Gallagher, describes a professional development (PD) program for teachers of mathematics who teach ELLs. Participants received instruction in second language acquisition as well as cultural and linguistic factors that may influence ELLs' learning, practices to support ELLs' learning of mathematics, growth mindset in the mathematics classroom, and mathematical knowledge for teaching rational numbers. The program included two summer institutes in subsequent summers, coaching visits, workshops during the school year, and a mathematics summer camp for ELLs. The authors suggest PD designers keep six principles in mind when supporting teachers of mathematics' development of productive beliefs, knowledge, and practices needed to effectively teach ELLs.

In Chapter 5, Alandeom W. Oliveira, Carla Meskill, and Sepideh Yasrebi describe how the practices of an algebra teacher evolved during her participation in a professional development program as she collaboratively developed, implemented, and reflectively assessed a series of three algebra lessons with a specialist in second language acquisition. Teachers' observations, collaborations, and reflections were coded, categorized, and presented in the chapter. They show how the professional development program impacted teacher's practices related to (re)structuring classroom activity, small-group work, and increased attention to language. The authors make a case for the use of complementary expertise as a form of professional development. They conclude that effectively teaching algebra to ELLs requires understanding what it means to "speak algebra", and recognition that even the most abstract algebraic concept can be taken up as talk and thereby learned.

Chapter 6, by Sarah A. Roberts, describes a professional development experience organized around a framework of questioning and four principles of effective mathematics instruction for multilingual students. Three seventh grade mathematics teachers participated in semester-long professional development with this framework grounding their work. This chapter provides four illustrations of practice, one for each principle, from the professional development, as the teachers considered how to prepare to implement these principles and questioning in their own practice. This work draws on the teachers' discussions in the professional development, focusing on their considerations for how to support multilingual students.

In Chapter 7, Gretchen P. Oliver and Karen M. Gregory focus on students' development of discipline-specific thinking competencies which demonstrate not only their conceptual understandings, but also their ability to comprehend and use academic language. This chapter provides three explicit examples of how mathematics teachers engaged high school ELLs in mathematical learning through such activities and shifted their pedagogy to provide them with equitable learning opportunities. Through meaning-making activities, teachers can facilitate ELL students' engagement with mathematical practices, paying simultaneous attention to language development and conceptual understandings.

Chapter 8, by Holly Hansen-Thomas and Juliet Langman, highlights the Functional Language Awareness approach, or FLA, which centers on examining and analyzing how mathematics teachers for Emergent Bilinguals (EBs) attend to and incorporate language into their teaching. The FLA approach puts *activity*, or what is happening in the classroom, at the core of the model. Activity is surrounded by language *functions*, *features*, and *contexts* that work collaboratively to help teachers think about language as a teaching tool in the mathematics classroom. As a conceptual tool, this approach assists teachers in creating student participation through language in the mathematics classroom. Drawing on discourse from both middle and high school mathematics classrooms, this chapter reveals how teachers of EBs use language within this approach to reach all learners, as well as to promote their participation, interaction, and language use in mathematics class.

Chapter 9, by Angela R. Thompson, Michelle Hale, and Alexander Radosavljevic, explores one teacher's response to her school district's demand that she teaches her bilingual students with project-based learning. Through action research, the authors investigate how the students engaged in the CCSS-M practice standards while also engaging in multi-language discourse, engineering, and mathematics by building model roller coasters. The authors conclude with guidelines for planning and assessment of mathematics projects as well as their insights and suggestions for doing mathematical modeling and discourse in a classroom where the students are bilingual but the teacher may not be.

In Chapter 10, by Amanda E. Lowry and Melinda Eichhorn, report on a case study exploring the way two in-service high school mathematics teachers have reflected upon and adjusted their practice, based on education and professional development, previous teaching experience, and self-initiated research for teaching mathematics to newcomers and ELLs. The authors conclude that teacher education programs and school districts can help address challenges math teachers of ELLs face by providing the PD and instruction, specifically focusing on how to support students in learning math while being supportive of their emotional and mental health needs, making math content more accessible to students, and supporting language acquisition for high school ELLs while addressing the mathematics content.

Chapter 11, by Susan Holloway and Gerardo De La Garza, provides a case study of in-service mathematics teachers' work at the secondary level and details learning experiences that focus on mathematics and language for ELLs, with an emphasis on teacher knowledge and practice. It describes the structures that the school has implemented for improving ELLs mathematics learning and teachers' use of recommended instructional strategies, activities, and assessment practices.

In Chapter 12, by Rachel S. G. Bower and Vanessa Z. Mari, presents an analysis of case studies of high school mathematics teachers working with an Emergent Bilingual (EB) population to examine written feedback and assessment. Studying assessment and teacher feedback allow a look at the reading, writing, and mathematical thinking being asked of EB students given both the content expectations and standards for mathematical practice of CCSSM. The in-depth analysis of the language used in written feedback provided for EBs by teachers addresses research recommendations for students' linguistic development and cultural needs.

The chapters in this book are clearly focused on the central theme of preparing secondary mathematics teachers to effectively teach ELLs. As a collective, these chapters address several of the recommendation for teacher education as called for in the consensus report on English Learners in STEM subjects (National Academies of Sciences,

Engineering, and Medicine, 2018). The increasing ELL population in schools in English-speaking countries around the world speaks to the urgency of developing secondary mathematics teachers' understanding of and effective responses to the issues associated with teaching ELLs. The integration of research and practice in the book is likely to appeal to a wider variety of educational stakeholders, including teacher educators, curriculum specialists, and instructional leaders.

References

de Oliveira, L. C. (2016). A language-based approach to content instruction (LACI) for English language learners: Examples from two elementary teachers. *International Multilingual Research Journal, 10*(3), 217–231.

de Oliveira, L. C., & Obenchain, K. (Eds.). (2018). *Teaching history and social studies to English language learners: Preparing pre-service and in-service teachers*. London: Palgrave Macmillan.

de Oliveira, L. C., & Shoffner, M. (Eds.). (2016). *Teaching English language arts to English language learners: Preparing pre-service and in-service teachers*. London: Palgrave Macmillan.

de Oliveira, L. C., & Wilcox, K. C. (Eds.). (2017). *Teaching science to English language learners: Preparing pre-service and in-service teachers*. London: Palgrave Macmillan.

Lucas, T., & Grinberg, J. (2008). Responding to the linguistic reality of mainstream classrooms: Preparing all teachers to teach English language learners. In M. Cochran-Smith, S. Feiman-Nemser, & D. J. McIntyre (Eds.), *Handbook of research on teacher education: Enduring questions in changing contexts* (3rd ed., pp. 606–636). New York, NY: Routledge.

Moschkovich, J. (2013). Principles and guidelines for equitable mathematics teaching practices and materials for English language learners. *Journal of Urban Mathematics Education, 6*(1), 45–57.

National Academies of Sciences, Engineering, and Medicine (NASEM). (2018). *English learners in STEM subjects: Transforming classrooms, schools, and lives*. Washington, DC: The National Academies Press.

National Center for Education Statistics. (2019). *English language learners in public schools*. Institute of Education Sciences. US Department of Education. Retrieved from https://nces.ed.gov/programs/coe/indicator_cgf.asp.

National Governors Association Center for Best Practices & Council of Chief State School Officers. (2010). *Common core state standards for English language arts and literacy in history/social studies, science, and technical subjects.* Washington, DC: Authors.

Schleppegrell, M. J. (2004). *The language of schooling: A functional linguistics perspective.* Mahwah, NJ: Erlbaum.

2

Preparing Mathematics Preservice Teachers for Teaching Emergent Bilinguals Through Concurrent Intervention

Ji Yeong I and Jiaqi Yu

Emergent Bilinguals (Garcia, Kleifgen, & Falchi, 2008) face dual challenges in mathematics classrooms: learning mathematics while developing fluency in the language of instruction. In the USA, Emergent Bilinguals (or multilinguals) in K-12 schools are generally called English language learners (ELLs) as English is the language of instruction in most public schools. However, we use Emergent Bilinguals (EBs) to reject other deficit-oriented terminologies (e.g., ELLs or LEPs) and to emphasize the fact that they already speak at least one language other than English. Although teaching EBs is becoming a necessary challenge for mathematics teachers, teacher preparation for effectively teaching EBs is far behind what is needed from the field. Indeed, in November 2014, over 30 states in the USA did not require teacher preparation programs to provide training in working with EBs (Education Commission of

J. Y. I (✉) · J. Yu
Iowa State University, Ames, IA, USA
e-mail: jiyeongi@iastate.edu

J. Yu
e-mail: jiaqiyu@iastate.edu

L. C. de Oliveira and M. Civil (eds.), *Teaching Mathematics to English Language Learners*, https://doi.org/10.1007/978-3-030-48355-5_2

the State, 2019). The insufficient teacher preparation regarding teaching EBs is concerning as the number of EBs continues to grow. Based on the current demand for teachers to meet the needs of EBs, mathematics teacher educators contend that teacher preparation programs should include rigorous EB or multicultural education (e.g., Durgunoglu & Hughes, 2010).

Furthermore, the current trends of mathematics education in the USA (e.g., Standards of Mathematical Practices of Common Core States Standards for Mathematics; National Governors Association Center for Best Practices & Council of Chief State School Officers, 2010), highlight the language demands involved in learning and applying mathematics. In addition, teachers should accommodate for the cultural and social factors associated with cognitive activities (Campbell, Adams, & Davis, 2007). The contexts embedded within mathematics problems often contain cultural aspects that are unfamiliar to EBs. Therefore, mathematics teachers must be aware that EBs may encounter unique challenges when engaging with class material, such as understanding the mathematical content and problem-solving aspects alongside any language demands and cultural differences. For example, mathematical tasks situated in the contexts of football or Thanksgiving holiday may provide EBs with additional difficulties to find an entry point. Without the ability to identify these challenges, teachers are not able to accurately assess EBs' understanding of mathematics, and in turn, cannot provide the appropriate support for them to learn mathematics. To address this challenge, we investigated what culturally responsive strategies preservice teachers (PSTs) consider when helping EBs make sense of cognitively demanding mathematical problems. These PSTs were receiving a series of interventions concerning their teaching approach while they worked with EBs and the teaching practices they implemented were analyzed using culturally responsive teaching framework, described in the next section.

Culturally Responsive Teaching in Mathematics

To teach EBs effectively, a sufficient body of literatures (e.g., Parker, Bartell, & Novak, 2015; Vomvoridi-Ivanović, 2011) employed or recommended culturally responsive teaching (Villegas & Lucas, 2002) or culturally relevant pedagogy (Ladson-Billings, 2014) and, more recently, culturally sustaining pedagogy (Alim & Paris, 2017). We adapted previous models of culturally responsive pedagogy (Aguirre & del Rosario Zavala, 2013; Hernandez, Morales, & Shroyer, 2013) in a way to sufficiently address the needs of EBs. By merging and refining the models, while focusing on EBs' needs, we came up with three categories of culturally responsive teaching strategies for EBs: *culture integration, cognitive demand,* and *linguistic support.*

Culture integration is aligned with a research-based EB strategy that incorporates a student's culture and prior experience into mathematics lessons (Chval & Chavez, 2011). As curriculum materials often include cultural contexts that EBs may not have prior knowledge about, teachers should be aware of these cultural aspects, and in turn, incorporate cultural experiences more familiar to EBs into their instruction to make meaningful connections between their home culture and the curricula.

Cognitive demand implies that teachers should value all students equally and provide them with an opportunity to learn rigorous mathematics. Assisting students to be critical and independent thinkers is essential in culturally responsive teaching. For example, teachers with low expectations may assume that EBs should not be given challenging tasks due to their English proficiency (Reeves, 2006), and as a result, they might ask EBs to complete only low-cognitive-demand tasks in mathematics (de Araujo, 2012). In contrast to this common misconception, researchers recommend that teachers should instead assign challenging mathematical tasks to EBs (Celedon-Pattichis & Ramirez, 2012; Moschkovich, 2010) and position EBs as competent problem solvers because inequitable positioning may result in fewer learning opportunities for EBs (Pinnow & Chval, 2014; Yoon, 2008).

Finally, *linguistic support* includes various non-verbal strategies that support EBs to overcome the language barrier that they may experience when solving a mathematical problem. Many alternative tools have been proposed and recommended to support the learning of mathematics among EBs, such as the application of physical activities (Ahn, I, & Wilson, 2011), gestures (Fernandes, 2012), visuals (I & Stanford, 2018), or manipulatives (Boakes, 2009). Moreover, while applying these alternatives to help EBs understand mathematics, teachers need to encourage them to participate in mathematical discussions because EBs are likely to have difficulties when explaining their mathematical ideas in English. Thus, teachers should offer and allow EBs to use multiple modes of communication, such as gestures, illustrating, or using their first language (L1). Salmona Madriñan (2014) found that students' first language serves as a tool to help them learn new content knowledge and their second language acquisition. Hence, rather than treating a student's first language as a deficiency, teachers should respect an EB's first language and culture and use them as intellectual resources (Moschkovich, 2007). As a student can use his/her first language to build a cognitive model of complex mathematical concepts before they are introduced to academic vocabulary, it is important to take advantage of their first language while helping them make sense of mathematic problems.

Context of the Study

We observed and analyzed three pairs of PSTs and EBs. The PSTs were White females who were pursuing middle-school mathematics licensure and enrolled in a university-based teacher preparation program located in the Midwest of USA. None of the PSTs had ever lived outside of the USA, had experience teaching EBs mathematics, or had ever received an EB-focused training. The PSTs were named Becky, Lucy, and Kate (pseudonyms). To build a one-on-one setting between PSTs and EBs, three middle-school EBs were recruited, two who were in 7th grade and one who was in 8th grade, to be part of the study. All of the EBs at that time had come to the USA from South Korea less than two years prior were identified as EBs and were enrolled in intermediate EB [ELL] classes by their school administration. For this study, we have named them

Se-Bum, Si-Young, and Ye-Jun (pseudonyms). Each PST was randomly assigned to work with one of the EBs. Once all the pairs were set up, each PST worked with the same EB over the course of five (once a week) sessions, so that they could use their earlier experiences with their paired student to design the following session. Although participating in this study was not required for any methods course, the PSTs volunteered for the sake of their own growth as educators.

During the five-week study period, all the PSTs were asked to develop a lesson plan based on cognitively demanding problems, which were chosen from the released items belonging to the following assessments: The Smarter Balanced Assessment Consortium, the Partnership for Assessment of Readiness for College and Careers, the National Assessment of Educational Progress, and the Program for International Student Assessment. The chosen problems satisfy the high-level cognitive demand criteria (Stein, Smith, Henningsen, & Silver, 2009). All pictures and representations were removed from each problem because one of our observation points of inquiry was to see how the PSTs' would modify the given problems through the addition of visuals.

Before the first meeting with all the PSTs—when the first intervention would be provided—each PST completed a pre-survey that was meant to elicit their beliefs regarding EBs. Once this first meeting was completed, the PSTs followed the same weekly routine as described in Fig. 2.1. First, the PSTs prepared a lesson plan based on a high-level cognitive demand mathematical problem (Stein et al., 2009), which was given to them a week ahead of time. They had the freedom to employ any instructional strategies they deemed appropriate and modify the problem in any way they believed was necessary. During a pre-interview, conducted individually and immediately before each PSTs' meeting with their respective EB, the PSTs were given an opportunity to explain their decisions in relation to their instructional plan. Afterward, the teaching session lasted about one hour, and the PSTs were interviewed after each teaching session so that they could reflect on their implementation. An intervention followed regarding research-based EB teaching strategies and the next problem was then given once the intervention was complete. Finally, the PSTs each submitted a written reflection, describing what they had learned from working with their EB and from the intervention. At the

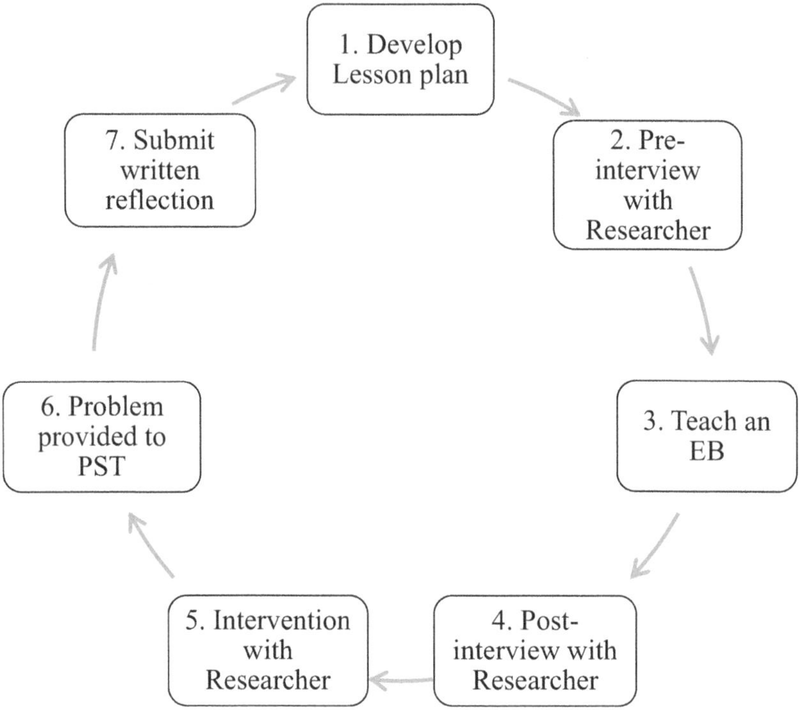

Fig. 2.1 The cycle of the PSTs' weekly routine and responsibilities

conclusion of the five sessions, each PST completed a post-survey and an individual exit-interview regarding the strategies they had implemented and any perceived changes they made to help their respective EB better understand mathematical concepts.

The intervention was designed to teach these PSTs research-based EB strategies (Chval & Chavez, 2011) and culturally responsive teaching (Villegas & Lucas, 2002). The first intervention session was given to the entire group before the PSTs met with their EBs. The session began with reading a real story of a Korean EB, an excerpt taken from Celedon-Pattichis and Ramirez (2012), to help the PSTs recognize the particular needs and culture of the EBs they were going to teach. The PSTs compiled a list of the difficulties EBs may encounter in English-only classes, shared their lists, and brainstormed how to help EBs overcome each of those difficulties. The other four interventions were given individually to each PSTs. The second intervention covered connecting mathematics to students' prior knowledge and experiences—which we also revisited in the fifth intervention with concrete examples. During these two intervention sessions, the PSTs were encouraged to use the EBs' life and cultural experience. Specifically, two versions of teacher-student dialogue were provided to help the PSTs identify and analyze the differences. Both dialogues illustrated the beginning of the same mathematics lesson, but one simply introduced a problem, while the other involved a conversation related to the student's life experience and connected the experience to mathematical concepts being taught. Other topics within the intervention sessions included how to utilize visuals by connecting mathematical information (I & Stanford, 2018) and building a rich environment for learning both mathematics and language. All interventions were administered using examples and hands-on activities rather than merely telling the PSTs how to teach mathematics.

Implementing Culturally Responsive Teaching Strategies

Our results are summarized below and divided into the three categories of culturally responsive strategies for teaching EBs: *culture integration, cognitive demand,* and *linguistic support.*

Culture Integration

We found two types of strategies used by the PSTs that fall within this category: integrating an EB's culture into instruction and assessing EB's understanding of cultural aspects embedded within problems. While integrating culture is a common strategy among all the PSTs, Lucy was the only one who often assessed the extent to which Si-Young understood a cultural factor embedded within problems. In contrast, Becky and Kate did not attempt to address any cultural factors in the problems although they frequently included their EBs' culture or life experiences in their lessons. The following excerpt is one example of Lucy and Si-Young's conversation before a problem was introduced:

Lucy: We are going to be working speed and transportation and time. Do you have a certain way of transportation you like? Like riding a car or…
Si-Young: … riding a car.
Lucy: Riding a car? What kind of cars do you like?
Si-Young: … Van?
Lucy: Vans? Is that what your parents have? Did you ride in a different way of transportation in South Korea?
Si-Young: No.
Lucy: Did you take buses or trains?
Si-Young: I rode trains and buses and a lot of things…
Lucy: So, a lot of things? What was your favorite?
Si-Young: Bus?
Lucy: Bus. Would you ride that a lot?
Si-Young: (Nodding)
Lucy: Well, could you see any similarities or differences from transportation here and South Korea?
Si-Young: In South Korea, I think it's closer, so we rode a lot of buses, but in America, they don't have a lot of buses.
Lucy: They don't? Yeah. I realize that too. They do have a lot of them on campus. So, I get to back and forth from my house through that. Do you have a lot of traffic in South Korea?
Si-Young: (Nodding)
Lucy: Yeah, so it might be easier to try a bus or a bike than just trying yourself. Cool. Well we're going to do some speed. Do you know what speed is?

In this excerpt, Lucy tapped into Si-Young's experiences in her home country multiple times, and Si-Young actively responded to these questions. Through the conversation, Lucy bridged the problem to her student's culture and, at the end of the excerpt as well as afterward, built a connection between Si-Young's experience and the mathematical problem about speed when she showed the toy speed limit sign and asked real-life related questions. After this conversation, Lucy asked Si-Young to designate certain locations to determine a speed given time and distance to teach the linear relationship between them and the concept of speed.

Moreover, when the PSTs attempted to integrate EB's culture or interests, it sometimes increased the students' confusion because they did not carefully assess the EB's understanding beforehand. For example, Ye-Jun's confusion arose when Kate changed the context of week 4's problem—designing a one-story house floor plan—to building an arcade in a basement because she heard Ye-Jun mention how he liked computer games. The following excerpt highlights the moment when she introduced this redesigned task:

Kate: First let me ask you, do you like video games?
Ye-Jun: Yeah.
Kate: What's your favorite game? Sports video game? Do you like computer? Like other games?
Ye-Jun: Sports game.
Kate: Sports game? Have you played any new one?
Ye-Jun: No.
Kate: What kind of video gaming system you have? You have X-box, PS 3?
Ye-Jun: (Shakes his head) Computer.
Kate: Computer? That's good, too. Um. today, we're going to build our own arcade. That's what we're going to be doing. Have you been to arcade before?
Ye-Jun: (silent for a while) No.
Kate: It's like, sometimes you go to pizza places, you go bowling alley, there's a separate area, that has all the video games you can play. They usually have a snack bar. You can win a prize.
Ye-Jun: No.
Kate: We're going to build an arcade today. So, let's look at what we have here. Do you want to read this to me?
Ye-Jun: (struggles to read the first word) The... Nugin?
Kate: Nguyens
Ye-Jun: What?
Kate: The name? [points "Nguyens"]
Ye-Jun: Yeah.
Kate: Nguyens.

Ye-Jun: Nguyens, are going to build an arcade in their basement. The floor of the arcade will be rectangular with a length of 30 feet and a width of 20 feet. [keep reading the task]

According to Kate's lesson plan, her intention was to make this problem engaging for Ye-Jun by building it around an interesting context to him. However, Ye-Jun did not know what an arcade was. He expressed his uncertainty by saying "no," but Kate neither recognized his confusion and nor provided additional aids, such as a photo of an arcade or a video. In addition, Kate included a Vietnamese name, Nguyen, in the modified task assuming Ye-Jun might be familiar with this name because he identified as Asian. However, he was not familiar with this Vietnamese name and struggled when reading it. Although Kate implemented the strategies with an intention to engage Ye-Jun's culture and interests, the modified context was not helpful because her strategies were not built on Ye-Jun's experiences. This is evidence that assumptions about EBs made without assessing their actual prior knowledge may cause confusion and obstacles than access for these students when solving mathematical problems.

Furthermore, we found that the PSTs did not attempt to incorporate any cultural or life experiences into their instruction during week 1 but did begin to implement them once they went through the relevant intervention. Notably, after the intervention, Becky made cultural connections more frequently than she had before. Lucy, who often referred to her student's cultural experiences throughout all five weeks of study, mentioned in her post-survey responses that the most effective strategy she learned from the intervention was cultural integration:

> I found it useful to connect my student's life experiences with mathematics. This helped her understand the relevance of the material and got her to notice mathematical happenings in her everyday life. For example, when I told her about speed she couldn't understand it, but then once I showed her a speed limit sign she recognized what the math formula meant.

Cognitive Demand

Our focus in this category was how the PSTs maintain or increase the cognitive demand of the problems. The common teaching practices with high-level cognitive demand include asking students to explain about their solutions and seek multiple solutions. Lucy made more attempts to apply these strategies than the other PSTs, yet, overall the PSTs made few attempts to use the strategy of *seeking multiple solutions* than *pressing for explanation*. The PSTs seemed to be satisfied when—after the EB struggled to communicate for an extended period—they heard one correct answer. Another reason the PSTs did not press multiple solutions could be the fact that they simply had not prepared for more than one solution pathway. For example, here was the problem for week 1:

> Claire is filling bags with sand. All the bags are the same size. Each bag must weigh less than 50 pounds. One sand bag weighs 58 pounds, another sand bag weighs 41 pounds, and another sand bag weighs 53 pounds. Explain whether Claire can pour sand between sand bags so that the weight of each bag is less than 50 pounds.

This problem was designed for eighth-grade students when it was released by the Smarter Balanced Assessment Consortium, and its suggested solution was use of the concept of average: If the sum of the weight of three bags is greater than or equal to 150 pounds, it is impossible to make all the bags weigh less than 50 pounds each. However, none of the PSTs used this method; instead, they stuck to addition and subtraction as means to adjust the weight of each bag so that they would be equal to or less than 49 pounds without considering rational numbers between 49 and 50. After several tries, the PSTs made conclusions without valid generalizations and did not seek other solutions that might have maintained the cognitive demand of the problem. Similar to this, the PSTs sometimes reduced the cognitive demand of a given problem. For example, Becky removed a challenging component from the first week's problem when simplifying its language. Although the problem appeared easier and less confusing, she changed the "Explain" task into a "Yes/No" question, which requires much less cognitively demanding work than explanation. Similarly, Lucy developed an easy warm-up activity during week 1, as shown below in Fig. 2.2.

This task requires mastery of significantly lower level mathematical knowledge compared to Si-Young's grade level; in fact, adding and subtracting within 20 is a first grade standard (1.OA.C.6; CCSSM; National Governors Association Center for Best Practices, & Council of Chief State School Officers, 2010), and Si-Young was in an eighth-grade Algebra 1 class, which Lucy had been informed. However, Lucy started providing cognitively demanding tasks after she witnessed both Si-Young's accuracy and prompted computation skills. Similarly, there was a clear change among all the PSTs, such as fewer attempts of reducing cognitive demand, as they had more experience with their EBs and attended the intervention sessions. For example, after receiving the intervention explaining how to build a rich environment in language and mathematics, Becky's use of simple computations or repeating the same procedures dramatically decreased during her implementation. The strategies introduced to the PSTs included providing open-ended questions, using multiple solutions and representations, and prompting the EBs to explain or discuss their thinking prior to providing answers. Implementing these strategies might have helped the participants maintain the cognitive demand of the tasks.

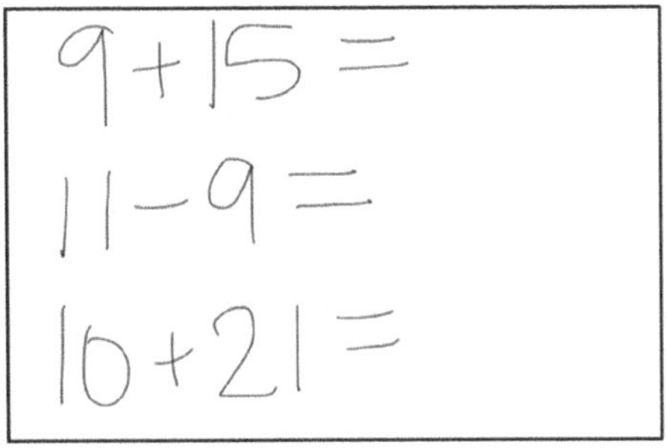

Fig. 2.2 Lucy's first mathematical activity to assess her student's mathematical knowledge

Linguistic Support

Various verbal and non-verbal methods to help the EBs understand the problems and express their mathematical thinking process were used by each PST. Visuals were the most commonly used method, while manipulatives and physical objects were employed less frequently than visuals.

All the PSTs consistently incorporated various visuals every week, such as charts, tables, pictures with captions, drawing on a grid paper, color coding, and post-its. In the beginning, the PSTs simply added context-related pictures, but the visuals applied during the later weeks connected to the mathematical information and situation of the problem (I & Stanford, 2018). Visuals were used even before the intervention addressed visual methods, although they were not always successfully implemented. For example, during week 2, Kate simply added one picture of a car, which related to the context of the problem but failed to convey any mathematical information. After the intervention addressed visuals, the PSTs applied visuals more appropriately for each mathematical situation. For instance, Kate included pictures of apartment floor plans connected to the information about each owner's living space, including captions that conveyed mathematical information. She purposefully chose these detailed visual examples to convey the different sizes of the units in the building. Similarly, Becky created a table to help Se-Bum, her assigned EB, comprehend how an apartment's price increased proportionally to the size. In the table, she included a picture to represent each word and used question marks to indicate the unknown information.

Visuals, gestures, and manipulatives were also used to provide multiple modes of communication between the PSTs and their EBs. The following conversation is an example depicting how Kate sought multiple ways to communicate with Ye-Jun.

Kate: So, one bag is going to weigh 42, one bag is going to weigh 56, and one bag is going to weigh 52 [she modified the numbers from the original problem].

Ye-Jun: (Draw three rectangles and enter 42, 56, and 52 each) This must be less than 50?

Kate: Yes.
Ye-Jun: (Write 7, 7, and 3 under each rectangle) Add this one here (draw an arrow from 7 under the box of 56 to 7 under the box of 42, and then write 49 and 49 under both boxes and 52 under the last box) it makes 49, 49, and this is 52. So, it's not possible.
Kate: You're right. It is impossible. Is there any other way we can possibly solve this problem? Besides getting the fourth bag.
Ye-Jun: Um… we add all this, (hand gesture including all three numbers) and it should be less than one hundred fifty, or [the] same.
Kate: You want to write that out for me? You want to write what you just thought of in your head?
Ye-Jun: Write words?
Kate: You can write it in numbers, you can write in words, just whatever you just did in your head. Can you show me?
Ye-Jun: (Draw 3 boxes and a big box underneath) Add these together (draw arrows from small boxes to the big box)
Kate: So you added those three numbers?
Ye-Jun: Yes.

First, Ye-Jun used his drawing to demonstrate his solution, which involved the concept of average. When Kate asked for a different method, he used a hand gesture to express his thinking, how he was adding all three numbers to see if the sum exceeded 150. Then, Kate asked him to write it out in numbers or words. Ye-Jun responded by drawing his idea. In this short instance, Kate and Ye-Jun communicated through four different modes of communication: drawing, gesturing, writing numbers, and writing words.

Discussions and Implications

The PSTs' use of strategies related to culturally responsive pedagogy evolved while they received specially designed interventions. The PSTs showed significant improvement in terms of implementing cognitive demand tasks, and the influence of the intervention was visible, especially in regard to *culture integration*. This is important because they started to use their EBs' culture as a resource (Moschkovich, 2010).

However, our results indicated that assessing EBs' cultural knowledge or explaining the cultural aspects were not frequently applied, although the attempts to integrate EBs' cultural components into instructions commonly appeared.

Moreover, this study speaks to the needs of teacher preparation programs by introducing a model for cultivating PSTs' skills for teaching mathematics to EBs. Working one-on-one with EBs prior to their student teaching can help PSTs foster both tangible and specific knowledge about teaching EBs, as well as leverage their competences in working with diverse students. This preparation could be accomplished by concurrently providing interventions and teaching opportunities, enabling PSTs to apply the EB strategies from the interventions within a short period of time. In addition, fieldwork with EBs is crucial because such teaching experience can change PSTs' preconceived notions of EBs, as noted in Lucy's case. In her exit interview, Lucy explained why she had such low expectations of Si-Yeong: "I assume[d] she [Si-Young] knew math well, I guess her English is not good. So, doing math in English would be a difficult part." This statement reveals her belief that a student's English proficiency may affect his/her mathematics performance. However, Si-Young completed the computing tasks quickly and accurately, which were well beyond Lucy's original expectations. After their first session, Lucy admitted her mistake and completely shifted her plan and teaching approach during the following weeks, holding much higher expectations of Si-Young.

Although working with EBs could be an important supplement to teacher education program, and perhaps be the catalyst that changes PSTs' deficit views of EBs, merely providing fieldwork opportunities will not provoke effective learning (de Araujo, Smith, I, & Sakow, 2018). The concurrent interventions, though, may be the needed component that enables PSTs to maximize their learning. One PST even mentioned that immediately implementing EB strategies she learned during interventions was the most beneficial aspect of this project. To provide a pathway for quick implementation, working with one EB or a small group of EBs could expand PSTs' capability to teach diverse learners.

References

Aguirre, J. M., & del Rosario Zavala, M. (2013). Making culturally responsive mathematics teaching explicit: A lesson analysis tool. *Pedagogies: An International Journal, 8*(2), 163–190. https://doi.org/10.1080/1554480X.2013.768518.

Ahn, R., I, J. Y., & Wilson, R. T. (2011). Teaching mathematics to English language learners using Moses' Five-Step Approach. *Teaching for Excellence and Equity in Mathematics, 3*(1), 20–28.

Alim, H. S., & Paris, D. (2017). What is culturally sustaining pedagogy and why does it matter? In D. Paris & H. S. Alim (Eds.), *Culturally sustaining pedagogies: Teaching and learning for justice in a changing world* (pp. 1–21). New York: Teachers College Press.

Boakes, N. (2009). Origami instruction in the middle school mathematics classroom: Its impact on spatial visualization and geometry knowledge of students. *Research in Middle Level Education, 32*(7), 1–12. https://doi.org/10.1080/19404476.2009.11462060.

Campbell, A. E., Adams, V. M., & Davis, G. E. (2007). Cognitive demands and second-language learners: A framework for analyzing mathematics instructional contexts. *Mathematical Thinking and Learning, 9*(1), 3–30.

Celedon-Pattichis, S., & Ramirez, N. G. (Eds.). (2012). *Beyond good teaching: Advancing mathematics education for ELLs.* Reston, VA: National Council of Teachers of Mathematics.

Chval, K. B., & Chavez, O. (2011). Designing math lessons for English language learners. *Mathematics Teaching in the Middle School, 17*(5), 261–265.

de Araujo, Z. (2012). *Transferring demand: Secondary teachers' selection and enactment of mathematics tasks for English language learners.* Doctoral dissertation, University of Georgia.

de Araujo, Z., Smith, E., I, J. Y., & Sakow, M. (2018). Preservice teachers' strategies for teaching mathematics with English learners. In T. G. Bartell (Ed.), *Toward equity and social justice in mathematics education* (pp. 217–240). New York, NY: Springer International Publishing.

Durgunoglu, A., & Hughes, T. (2010). How prepared are the U. S. preservice teachers to teach English language learners? *International Journal of Teaching and Learning in Higher Education, 22*(1), 32–41.

Education Commission of the State. (2019). *What ELL training, if any, is required of general classroom teachers?* Retrieved from http://ecs.force.com/mbdata/mbquestNB2?rep=ELL1415.

Fernandes, A. (2012). Case 2: Attending to student gestures. In S. Celedon-Pattichis & N. G. Ramirez (Eds.), *Beyond good teaching: Advancing mathematics education for ELLs* (pp. 145–151). Reston, VA: National Council of Teachers of Mathematics.

Garcia, O., Kleifgen, J. A., & Falchi, L. (2008). *From English language learners to emergent bilinguals. Equity matters. Research review no. 1.* Retrieved from http://eric.ed.gov/?id=ED524002.

Hernandez, C. M., Morales, A. R., & Shroyer, M. G. (2013). The development of a model of culturally responsive science and mathematics teaching. *Cultural Studies of Science Education, 8*(4), 803–820. https://doi.org/10.1007/s11422-013-9544-1.

I, J. Y., & Stanford, J. (2018). Preservice teachers' mathematical visual implementation for emergent bilinguals. *Mathematics Teacher Educator, 7*(1), 8–33. https://doi.org/10.5951/mathteaceduc.7.1.0008.

Ladson-Billings, G. (2014). Culturally relevant pedagogy 2.0: A.k.a. The remix. *Harvard Educational Review, 84*(1), 74–84.

Moschkovich, J. N. (2007). Bilingual mathematics learners: How views of language, bilingual learners, and mathematical communication affect instruction. In N. S. Nasir & P. Cobb (Eds.), *Improving Access to Mathematics*. New York and London: Teachers College Press.

Moschkovich, J. N. (Ed.). (2010). *Language and mathematics education: Multiple perspectives and directions for research.* Charlotte, NC: Information Age Pub.

National Governors Association Center for Best Practices, & Council of Chief State School Officers. (2010). *Common core state standards for mathematics.* Washington, DC: National Governors Association Center for Best Practices, Council of Chief State School Officers.

Parker, F., Bartell, T. G., & Novak, J. D. (2015). Developing culturally responsive mathematics teachers: Secondary teachers' evolving conceptions of knowing students. *Journal of Mathematics Teacher Education*, 1–23. https://doi.org/10.1007/s10857-015-9328-5.

Pinnow, R. J., & Chval, K. B. (2014). Positioning ELLs to develop academic, communicative, and social competencies in mathematics. In *The Common Core State Standards Serie. Common Core State Standards in Mathematics for English language learners* (pp. 21–33). Alexandria, VA: TESOL International Association.

Reeves, J. (2006). Secondary teacher attitudes toward including English-language learners in mainstream classrooms. *The Journal of Educational Research, 99*(3), 131–142.

Salmona Madriñan, M. (2014). The use of first language in the second-language classroom: A support for second language acquisition. *GIST Education and Learning Research Journal*. Retrieved from https://eric.ed.gov/?id=EJ1062663.

Stein, M. K., Smith, M. S., Henningsen, M. A., & Silver, E. A. (2009). *Implementing standards-based mathematics instruction: A casebook for professional development* (2nd ed.). New York: Teachers College Press.

Villegas, A. M., & Lucas, T. (2002). Preparing culturally responsive teachers rethinking the curriculum. *Journal of Teacher Education, 53*(1), 20–32. https://doi.org/10.1177/0022487102053001003.

Vomvoridi-Ivanović, E. (2011). Using culture as a resource in mathematics: The case of four Mexican-American prospective teachers in a bilingual after-school program. *Journal of Mathematics Teacher Education, 15*(1), 53–66. https://doi.org/10.1007/s10857-011-9201-0.

Yoon, B. (2008). Uninvited guests: The influence of teachers' roles and pedagogies on the positioning of English language learners in the regular classroom. *American Educational Research Journal, 45*(2), 495–522.

3

Redesigning a Statistics Literacy Course to Help Prepare Pre-service Teachers to Teach (Statistics to) English Language Learners

Lawrence M. Lesser

The Nature of Statistics

Statistics is a domain that uses mathematics, but is its own science, not a branch of mathematics (Cobb & Moore, 1997; Moore, 1988). And statistics education has established (apart from mathematics education) its own organizations, conferences, and journals as well. Lesser and Winsor (2009, p. 6) review deep differences such as "observational and experimental data have very different interpretations but use the same mathematical models of statistical theory" and "different foundational conceptions of probability or inference play out in teaching and research today (in a way that does not occur in mathematics)." It is often not sufficiently recognized how data are "numbers with a context" (Cobb & Moore, 1997), how the context of a question can determine the type of variable (Lesser, 2013a), and how context affects qualitative features of

L. M. Lesser (✉)
The University of Texas at El Paso, El Paso, TX, USA
e-mail: Lesser@utep.edu

L. C. de Oliveira and M. Civil (eds.), *Teaching Mathematics to English Language Learners*, https://doi.org/10.1007/978-3-030-48355-5_3

quantitative analysis (Gephart, Jr., 1988; Huberty, 2000) and interpretation of *p*-values (Betensky, 2019) or correlations (Akoglu, 2018). As English language learners (ELLs) researchers know, context is also crucial in deducing the meaning of words and in motivating student learning, so it is natural and important to explore ELLs in statistics.

When redesign of this course began, there was much research literature on teaching mathematics to ELLs, but virtually none specifically in statistics courses before Lesser and Winsor (2009). Thus, there was a need for careful investigation without just assuming that dynamics and findings from math contexts would all transfer. And even when tools could transfer, thoughtful reflection was needed to identify the best examples and counterparts. With deep motivation and empathy (Lesser, 2015), these years of exploration were greatly helped by working with a bilingual colleague (who had experience teaching math to ELLs and doing scholarship in this area) and later having regular meetings and trainings with a cohort of faculty colleagues in the LEAP-UP grant (see Acknowledgements). After describing the context for the redesign, this chapter will share the strategies that proved to be useful in that context and that can also help inform instructional and curricular decisions by instructors in other settings.

Context for the Course

The course was redesigned at The University of Texas at El Paso (UTEP), an R1 university on the US-México border, and about 5% of its students commute from México. UTEP's student population is about 50% first-generation, and—like the region—is about 80% Hispanic.

The multi-section course the author redesigned, coordinates, and often teaches is a statistical literacy course required for some majors (e.g., BA in applied learning and development, BA/BS in psychology) and is an option to satisfy requirements in other programs (BSW in social work, pre-health professions undergraduate certificate). The redesigned course selected Jessica Utts' *Seeing through Statistics*, a textbook which develops a concept with examples, numbers, and words, before any formulas. This sequence makes chapters less daunting and more likely to actually be read by students. Also, the sequence nicely parallels the common ELL-friendly technique of using informal language before academic vocabulary and syntax.

During its redesign, the course became part of the University Core Curriculum, which necessitated meeting these (slightly paraphrased) domain-specific educational objectives specified by the Texas Higher Education Coordinating Board: (1) apply statistical methods to modeling and real-world situations, (2) represent and evaluate information in multiple ways, (3) expand reasoning skills to develop convincing arguments, (4) use appropriate technology to enhance understanding, solve problems and judge reasonableness of results, (5) interpret models and draw inferences from them, (6) recognize the limitations of models, and (7) recognize evolving and interdisciplinary nature of the discipline. While this course (by being housed in a department of mathematical sciences rather than in a college of education) is officially viewed as a content (not pedagogy) class, pedagogy was nevertheless modeled and pedagogical content knowledge incorporated as the content was taught.

Audience

The audience for this chapter is instructors assigned to teach (or prepare students to teach) statistics, though the instructors may have minimal statistics experience or coursework. Indeed, statistics may be the subject most often taught by someone without work experience or a degree in it. The audience of the course includes students who are or were ELLs. When the redesign began, some 80% of course enrollees were pre-service teachers, almost all of whom end up teaching in this geographic region, a region where a high percentage of schoolchildren are ELLs, thus yielding a "closed loop" to amplify impact of the course. While this chapter focuses on the teacher candidates (TCs), we note that the pedagogical strategies were appreciated by all students in the class.

Many TC-oriented additions to the course were made, including adopting the textbook mentioned earlier, and making connections to children's literature (e.g., Smith, 2002), national content standards (e.g., Franklin et al., 2007; NCTM, 2000), state content and English language proficiency standards, state teacher certification tests, pedagogical content articles (e.g., from *Statistics Teacher Network* or *Teaching Statistics*), technology, manipulatives (e.g., a quincunx as well as varied dice, spinners, cards, coins, etc.), and a data collection project (Lesser, 2008). In some semesters, students were also offered (as an alternative to the data project requirement) an option to write a statistics K-12 data-based lesson plan (accommodating learners at each level of Texas' English language proficiency standards, following chapter 74.4 of the Texas Administrative Code) or write a reflection paper on a 20-hour service-learning experience arranged in a local public school setting (Lesser, 2008). Frequent asides ("How do you think a fifth-grader might interpret this?") were made to build TCs' capacity to teach while discreetly consolidating their own knowledge as well. Rationale for pedagogical choices was frequently shared, such as giving and discussing a quiz based on a study (Brown & Ryoo, 2008) on using the everyday-language-first strategy on mostly Latinx fifth-graders.

Strategies Used and Modeled

Word-Level Strategies

One word-level strategy is to call attention to words (see Table 3.1) that may have only slightly different spelling or pronunciation. Irazábal (2018) describes motivation and strategies for helping ELLs with correct pronunciation and the northeast quadrant of Table 3.1 lists homograph words often encountered in statistics, where syllabic stress aligns with the part of speech being used. The northwest quadrant illustrates lexical ambiguity where the same word has different meanings in different registers (even between statistics and mathematics) and everyday language can interfere when meaning is different. This is the most challenging situation for students and requires the most explicit attention and discussion.

While it is not unusual to encounter words whose meaning varies among contexts, it is much less recognized that a word or a symbol can have more than one meaning within the same context, such as within

Table 3.1 Words with similar pronunciation and/or spelling

	Same pronunciation (but different meaning among registers)	Different pronunciation
Same spelling	Bias, correlation, discrimination, median, normal, parameter, power, random, range, significant, skew, variation (Kaplan, Fisher, & Rogness, 2009; Lesser, 2010b; Lesser & Winsor, 2009; Wagler & Lesser, 2011)	Accent depends on part of speech: affect, aggregate, approximate, attribute, compound, contrast, coordinate, decrease, estimate, increase, record, reject, subject, survey (Lesser, Wagler, Esquinca, & Valenzuela, 2013, p. 10)
Different spelling	Discrete/discreet; compliment/complement (Lesser & Winsor, 2009)	Causal/casual; median/medium (see Table 1 of Lesser & Winsor, 2009)

the same statistics textbook (Lesser & Winsor, 2009). For example, *bias* can refer either to an individual's personal prejudice or to a measurement process' systematic error. The symbol N may denote population size in one chapter and a normal distribution in another. The Greek letter β may denote a Type II error probability in one chapter and a regression slope coefficient in another. The symbol π may in one chapter denote an unknown population proportion (between 0 and 1) and in another chapter be the mathematical constant (3.14159....) in the normal distribution pdf.

A common word-level technique is to recognize words that are *cognates*. Spanish statistical terms at http://isi.cbs.nl/glossary/bloksp00.htm indeed include many cognates such as: bias, bloque, causalidad, coeficiente de correlación, condicional, control, dispersión, distribución normal, error, extrapolación, hipótesis nula, independencia, mediana, moda, multivariable, percentil, precisión, probabilidad, regresión lineal, residual, tabla de contingencia, tendencia central, unimodal, variable predictiva, and variancia. Another role cognates can play is that a Spanish-speaking ELL is more likely than a monolingual native English speaker to infer the meaning of words that are much less common in the everyday English register than are their Spanish cognates in the everyday Spanish register (Lesser et al., 2013). For example, the common everyday

Spanish word *confundido* helps recognize the less common English word *confounded* as confused or "all mixed up," which supports the statistical meaning: If X and Z are confounded in predicting Y, we can't separate out what effects on Y are due to X and what are due to Z. Making connections such as these not only modeled helping ELL students, but also demonstrated to all students how knowing another language is a resource and an asset.

There are resources with varying levels of word-level support teachers can offer during class. Teachers who want to give students access to an English-Spanish glossary of mathematics terms that includes definition and illustration of each concept can keep on hand a book such as COMAP (2004). Before starting an exam, students were explicitly told they could quietly approach and ask risk-free about wording, with the instructor happy to answer any question as long as it was not the essential point being tested. After a student mentioned during an exam that she could answer a question if she simply could be given the Arabic word for a particular statistics term, the instructor was inspired to give students access to a website (http://isi.cbs.nl/glossary) of terms in 29 languages, thus offering (without giving away definitions) ELLs easier access to content they may have learned in their first language.

Specific resources were incorporated into the syllabus and/or learning management system to facilitate content connections from other languages. Students were introduced to collections of applets (e.g., http://nlvm.usu.edu/es/nav or http://eduteka.icesi.edu.co/mi/) to explore graphical representations, regression models, and probability concepts in (or toggle between English and) another language. Interestingly, Viera (2019) found that while ELLs who used such resources found them helpful, there were also ELLs who avoided such resources so they would feel more ready for future material (and tests) in English. Well-known ELL tools for learning vocabulary are in Lesser (2011a), such as a *word square* illustrated for the statistical mean.

Phrase-Level Strategies

Statistics also includes distinctive language patterns (e.g., the double negative of "fail to reject the null") and idiomatic intact phrases (e.g., "in the long run") that need to be made explicit to students. A classroom practice of colleague Amy Wagler was adopted, using quotation marks to signal such phrases (Lesser & Winsor, 2009) which may not yield to word-by-word translation, such as sum of squares, regression to the mean, in the long run.

Table 3.2 gives examples of many ways that can be used—at least initially—to keep idiomatic or less intuitive phrasing from making concepts inaccessible. Uncited entries are not original but lack a clear best citation. Many of these phrasings may especially help culturally and linguistically diverse learners use their everyday English to make a bridge to academic language.

Table 3.2 (Initial) alternatives to statistics textbook phrasing

Common textbook language	Alternative wording
Sampling distribution of $\bar{x}$	Histogram of sample means
Histogram	Quantitative bar graph (Rumsey, 2009)
Z transformation	Change to standardized units
Type I error	False alarm
Type II error	Missed detection
Margin of error	Margin of variability (Rumsey, 2009)
Confidence coefficient	Number of standard errors
Expected value	Average (mean) value [of a random variable]
In the long run	Long-term proportion (Sullivan, 2010); in the long term (Lesser, Wagler, & Salazar, 2016)
Longest run of heads	Largest number of consecutive heads (Lesser et al., 2016)
Unimodal, right-skewed distribution	A hill with a long tail to the right [or a drawing or gesture] (Lesser & Winsor, 2009)
Run an experiment	Conduct/do an experiment
Disjoint; mutually exclusive	Nonoverlapping
P(A or B)	P(A or B or both) (Rumsey, 2009)

And yet, it is important to note that informal language can have pitfalls, such as how students may unduly fixate on a histogram's "bumpiness" when instructors use the word "spread" instead of "variability" (Kaplan, Rogness, & Fisher, 2012). Also, using the word "average" (or "center") is ultimately insufficient because we don't know whether it denotes, for example, the median or mean. (Not to mention that the math register has several means: arithmetic, geometric, harmonic, etc.)

Lesser et al. (2016) identified helpful instances where Spanish adds much clarity to idiomatic English phrases in probability. For example, each of the four words in the common textbook phrase *in the long run* is one of the 1000 most common English words and yet the meaning of this intact phrase is idiomatic. But the Spanish phrase *en el largo plazo* translates to *in the long term*, which is clearer than *in the long run*. Also, the Spanish phrase *el mayor número de caras sucesivas* translates to "the largest number of successive [or consecutive] heads," which is clearer than the commonly used idiomatic phrase *longest run*. Once again, this demonstrated to all students how Spanish is a resource.

Sentence/Paragraph-Level Strategies

The course redesign included identifying specific language patterns in statistics that require attention to structures, not just individual words. For example, consider sentences that use conjunctions. The elegant solution of Fernandes, Anhalt, and Civil (2009) of deleting "if" from a mathematics word problem to avoid the everyday connotation of "if" as "implausible" turned out to transfer well to the frequent "if" phrases in statistics (e.g., "if a sample is random," "if a distribution is normal," "if variables (or events) are independent," "if a *p*-value exceeds .05," etc.), changing "If events A and B are disjoint, find the probability that…" to "Events A and B are disjoint. Find….." Another example of a challenging conjunction is "or" because it is usually "exclusive" (i.e., excluding overlap) in its everyday use (and even in some mathematical scenarios, such as specifying the solution set for $|x| > 5$ or for $x^2 = 4$), but inclusive in mathematical set theory and probability.

Well-known ELL tools for generating sentences such as a *sentence frame* were applied to the definition of a *z*-score and to the reporting of the result of a hypothesis test (Wagler & Lesser, 2011). Students were shown and encouraged to explore online resources that supported language development, such as a grammar parser http://www.statlit.org/GC/T2/SLRSV.aspx (Burnham & Schield, 2005) to get feedback on their sentences attempting to express the correct part-to-whole relationship when given tables or graphs.

Conceptual understanding was also assessed and developed by giving students occasional short writing assignments. A particularly illuminating example happened when students were asked to name the explanatory and response variables in (and discuss the ethics of) a disguised baby experiment (Begley, 2009) where "scientists dressed newborns in gender-neutral clothes and misled adults about their sex" and had adults describe the baby's emotional state. Many students stated "babies" as the explanatory variable instead of something like "stated gender of baby," which led to developing an assessment on identifying variables. And noticing that students often treated "males" and "females" as two quantitative variables instead of as two possible values of a single categorical variable ("gender"), a metacognitive question was developed ("When you ask the first person in your sample the associated question, what will you write down on your datasheet?") that students could ask themselves to determine if a given variable was categorical or measurement (Lesser, 2013a). Another example of a writing assignment used related to how the media depicts correlation or causation (http://jfmueller.faculty.noctrl.edu/100/correlation_or_causation.htm).

A helpful way to promote and model discourse was the adoption of the See-Think-Wonder thinking routine (which could be viewed as a kind of sentence starter) to improve (and make more visible) students' statistical thinking, as illustrated in Pfannkuch, Regan, Wild, and Horton (2010). And Lesser and Kephart (2011) illustrate in a statistics context the role of discourse features such as varying participation structures and having the instructor positioned as a facilitator and co-learner.

In addition to these strategies used in front of students, there were also strategies identified for use behind the scenes to make instructional materials and assessments more accessible for all students, especially

ELLs. Instructors can improve readability by increasing lexical simplicity (e.g., using more common words), grammatical simplicity, cohesion, and narrativity and to assess that improvement using readily available tools such as LexTutor, Coh-Metrix, and Microsoft Word (Lesser & Wagler, 2016).

Culturally Relevant Context

An instructor needs to explicitly ensure and not just assume that students have the background to understand the real-world context for a statistical example. For example, it was unrealistic in hindsight to expect students from a high-poverty desert community to relate to a skiing context and not every ELL student even recognized the word *ski* (Lesser & Winsor, 2009).

With other examples, ELLs or immigrants may recognize the context but not realize how it differs in the USA. For example, statistics textbooks often use the judicial analogy where the null hypothesis is "innocent unless proven guilty," a stance which is not consistently implemented in some countries, and ELLs had a different response pattern ($p = 0.006$) to that part of the analogy (Lesser, 2010b). Also, instructors need to mention explicitly the traits and associated language of classroom probability manipulatives (e.g., dice, cards, coins). For example, an ELL college student did not recognize *tails* as the name of a side of a (US) coin (Lesser & Winsor, 2009).

Other potential cultural pitfalls include assuming students are comfortable with US contexts involving sports, gambling, stock market, or polls about topics that are taboo in highly conservative cultures (Abdelbasit, 2010). But culturally relevant pedagogy is not just about avoiding pitfalls—it is also a resource that can be used mindfully to engage and connect with students. For example, the class enjoyed rich discussions that connected content to examples from Latinx culture, such as Toma Todo (Lesser, 2010c) and la lotería (Lesser, 2013b). And the presentation of data collection projects during the last few class meetings is always a huge highlight because each team is presenting on a question and context that they chose as meaningful to them. (This is one of several reasons why the redesign replaced an in-class exam with a project.)

Thinking Like a Statistician

An important part of preparing TCs with knowledge and experiences goes beyond content topics to include attention to the process and practices of doing statistics so that they experience on some scale what it means to talk, write, think, and act like a statistician. Some of this is motivated by process standards. Franklin et al. (2007) give a framework for statistical problem solving as an "investigative process" with the components of question formulation, data collection and analysis, and interpretation of results. Recommendations of the GAISE College Report ASA Revision Committee (2016, p. 3) include: "integrate real data with a context and purpose" and "use technology to explore concepts and analyze data." The data collection project is a good example of how these recommendations were implemented.

While Texas is not a Common Core state, the course objectives supported Core mathematical practices to reason quantitatively, critique reasoning of others (e.g., critiquing how statistics is presented by advertisers, activists, journalists, etc.), use appropriate tools strategically (the course included a session in a computer lab to learn how to use Excel and Minitab for their data collection projects, etc.), and attend to precision (e.g., margin of error concept, labeling units and axes of a graph). Specific activities and interventions (e.g., Lesser, 2018) were often explicitly aligned to Common Core. That said, statistics educators have identified areas where the Common Core needs further development. For example, the thoughtful analysis by Groth (2019) of the Common Core Math Standards recommends that, in the absence of explicit learning activities for grades 4–5, teachers of those grades will need to help students question their natural tendencies to focus extensively on the mode when summarizing data.

Like many states, Texas has its own College and Career Readiness Standards, and the redesigned course covers the standards for probabilistic reasoning (by supplementing the course textbook with more development of sample space) and statistical reasoning (by implementing the data collection project, etc.).

A pedagogy that included open-endedness and spontaneity was cultivated to help address the critique (of science education, but applicable to many statistics courses) by Willis (1995, p. 1): "Hands-on experiences have typically been limited to 'cookbook labs' with outcomes known in advance, not real investigations. In most classrooms, science has been presented as an inert body of knowledge to be assimilated, rather than a process of inquiry and a way to make sense of our world." The instructor would therefore often "think out loud" with new or atypical scenarios and give students the opportunity to explore and discuss problems that had more open-endedness than typical mathematics exercises, such as the scenarios of average class size and Simpson's paradox in Lesser and Kephart (2011). Also, anonymous data collection demonstrations occurred in class that required judgment to interpret, such as when the variable "hours of sleep per week" yielded outlier values such as 7 or 20. The class also experienced probability experiments with asymmetrical objects such as rounded-back thumbtacks (Lesser, 2018) whose probabilities did not follow the usual uniform sample space.

Another way to support more authentic engagement "as a statistician" is to offer sustained engagement with a real-world context (Brown, 2019), rather than the typical mathematics-style serving of simplified word problems with one task each. One take-home assignment for the most recent semester's students was to read about the Flint, Michigan water crisis and go through a progression of tasks and questions involving numerical and graphical representations of data as well as conceptual foundations (Loux & Gibson, 2019). The author also shared examples of his own past work as a state agency statistician by discussing statistical and nonstatistical issues that arose in the application of Census data to the state's decennial redistricting process (Rubel, Driskill, & Lesser, 2012).

The most important and sustained way students experienced thinking like a statistician is when they did the data collection project. On a small scale, students had to follow all the basic steps of research, starting by submitting a proposal (as human subjects researchers must do with an IRB) in which they specified details, including: research question/topic they those, nature and source of data to be collected, mode of collection, exact wording of questions, and mechanism to ensure anonymity or confidentiality. The explicit process forced many students to fully understand, for example, how hard random selection of participants is to implement when there is no obvious sampling frame. The instructor gave feedback on the proposal and then checked one of three categories: Project approved as written, project approved if a specified change is made, or re-submission of proposal needed to address major specified concerns. The design of this assessment was informed by literature such as Short and Pigeon (1998), Halvorsen and Moore (2005), and Roberts (1992). This aspect is arguably one of the most important in the course in light of the observation by Joiner (1988, p. 53): "Most people who teach statistics have never practiced statistics. And even those who have practiced statistics seldom use it in their own lives. Most students get advanced degrees without ever having designed, carried out, and analyzed an experiment or something of importance to them. Thus few teachers and fewer students know what it is like to gather and analyze useful data." Students also were encouraged to attend an on-campus research symposium in which undergraduate or graduate students presented their research in a gallery session of posters that TCs could browse and discuss statistical elements that they recognized. In a reflection, one TC wrote: "I could definitely see myself presenting one of those posters sometime in my near educational future. My dream is to become an early childhood bilingual educator. I would really like to have the opportunity to research techniques, methods, activities, anything and everything that would prepare my bilingual students to succeed and to find different methods that would work better in my children's learning abilities."

Strategically timed to support the data analysis portion of the project, the course added a technology lab day where students met in a classroom where they could all readily access computers and used Excel and Minitab to experience a guided exploration of all major topics of the course, including being able to produce camera-ready numerical and graphical summaries of data. It should be noted, however, that the emphasis on these fairly user-friendly packages is an intentional deviation from what real-world statisticians do because the latter are more likely to use the more powerful packages SAS or R which would not at all be user-friendly enough for a terminal course for non-majors.

Near the end of one semester early in the redesign period, students in two author-taught sections were given an anonymous Likert survey to assess what aspects of the course gave them insight into "how statisticians work or think." The 57 respondents (representing a 86.4% response rate) gave highest ratings to connections to the field of education, use of technology, and in-class data collection and lowest (but still quite positive) ratings to the textbook itself and the instructor's sharing of his past experiences as a statistician. In hindsight, this is not surprising because the TCs are being most responsive to the most immediate aspects. To the overall statement "This course would have been better if it had not included the above-mentioned ways to explore 'how statisticians work or think' and just stuck to lecturing/practicing basic math content," over 90% expressed some level of disagreement (with 71.9% of respondents strongly disagreeing).

Other Strategies

The usefulness of mnemonics in second-language vocabulary learning (e.g., Cohen, 1987) inspired creation, incorporation, and assessment of mnemonics in statistics to help recall certain facts or procedures, and free up students' cognitive resources for more higher-order thinking (Mocko, Lesser, Wagler, & Francis, 2017). Some mnemonics are letter/language-based while others are perhaps even more ELL-friendly by being visual in nature. Conceptual analogies (Lesser, 2016) were also incorporated, as well as concrete models such as representing the median with a lesser-known pulley-and-loop model (Lesser, Wagler, & Abormegah, 2014).

For many major course topics, interactive statistics songs (Lesser et al., 2019) were incorporated for which students could enter inputs (in response to prompts) and then hear the completed song with their inputs. The medium is ELL-friendly because students can pause, rewind, and replay them. Lyrics are delivered slower than speech and are more predictable and memorable due to cues of rhyme and rhythm and the use of direct conversational sentence structure with less jargon than textbook prose. And the pre-song questions are culturally inclusive by offering background on the real-world context or culture of the vignette. For example, where a prompt for the song "Central Limit Theorem" refers to a roulette wheel, there is a button that can be clicked on to obtain a definition and image. And the song "Hypothesis on Trial" makes sure the user realizes what a criminal trial's presupposition of the defendant is in the USA and offers a clickable hint for anyone unsure of the meaning of "convict" or "acquit." Completed songs were also used to intervene against misconceptions (Lesser, 2018).

The course also used ABCD classroom voting cards (http://bit.ly/USCOTS2019Lesser; Lesser 2011b) as a low-stakes way to encourage any student who may not seem readily able or willing to contribute to real-time in-class discussion. Also, the integration of diversity, ethics, gender equity, and social justice examples into the course (Lesser, 2007, 2008, 2010a, 2014) parallels the moral mandate to make content accessible and assessment fair for all students (Deygers, 2019).

Final Thoughts

While this chapter focuses on Spanish (since it's the first language of roughly 80% of US K-12 students who are ELLs), other languages can also be a resource in statistics, and this is yet another example of how this chapter's concepts can be applied to other cultural and linguistic environments. For example, the mean's conceptual interpretation as a "fair share" or "leveling value" is reflected in the Arabic origin of the word *average* (from the context of sharing the expense of damages to shipped cargo) and also in the Malay language, as in this exercise (COMAP, 2009, p. 177):

> In Malay, the expression for the mean is *sama rata*, which roughly translates as "same level". To understand this cultural and conceptual connection ... make stacks with 3, 7, and 8 chips. Redistribute chips among the stacks until they are at the same level and explain how this relates to the mean."

Lesser and Winsor (2009, p. 8) note: "Navigating the academic language of statistics is challenging also for students whose first language is English (Garaway, 1994; Nolan, 2002; Ortiz, Cañizares, Batanero, & Serrano, 2002; Rangecroft, 2002)." Indeed, survey research by Lesser et al. (2013, p. 17) found "ELLs and non-ELLs sometimes perceive or encounter statistical language the same way." And so, this chapter's pedagogical strategies help not only ELL students navigate the terrain of statistics, but (not unlike how many able-bodied individuals find curb cuts or closed captions helpful) *all* students.

Acknowledgements This work was supported in part by Josie Tinajero's US Department of Education grant Project LEAP-UP (T195N070132), as well as the author's National Science Foundation grant Project SMILES (DUE 1544237). The author also thanks the editors and peer reviewer for very helpful feedback.

References

Abdelbasit, K. M. (2010). Teaching statistics in a language other than the students'. In C. Reading (Ed.), *Proceedings of the Eighth International Conference on Teaching Statistics*. The Netherlands: International Statistical Institute.

Akoglu, H. (2018). User's guide to correlation coefficients. *Turkish Journal of Emergency Medicine, 18*(3), 91–93.

Begley, S. (2009). Pink brain, blue brain: Claims of sex differences fall apart. *Newsweek, 154*(11), 28.

Betensky, R. A. (2019). The *p*-value requires context, not a threshold. *The American Statistician, 73*(Suppl. 1), 115–117.

Brown, B. A., & Ryoo, K. (2008). Teaching science as a language: A "content-first" approach to science teaching. *Journal of Research in Science Teaching, 45*(5), 529–553.

Brown, K. M. (2019). More questions and fewer contexts: Designing exercises for statistics courses. *Journal of Statistics Education, 27*(3), 216–224. https://doi.org/10.1080/10691898.2019.1669508.

Burnham, T., & Schield, M. (2005). An online grammar-parsing program that decodes ordinary English descriptions and comparisons of percentages and rates. In *Proceedings of the Joint Statistical Meetings, Section on Statistical Education* (pp. 2202–2208).

Cobb, G. W., & Moore, D. S. (1997). Mathematics, statistics, and teaching. *American Mathematical Monthly, 104*(9), 801–823.

Cohen, A. D. (1987). The use of verbal and imagery mnemonics in second-language vocabulary learning. *Studies in Second Language Acquisition, 9*(1), 43–61.

Consortium for Mathematics and Its Applications (COMAP). (2004). *Mathematics as a second language: [English/Español] Mathematical glossary*. Bedford, MA: Author.

Consortium for Mathematics and Its Applications (COMAP). (2009). *For all practical purposes: Mathematical literacy in today's world* (8th ed.). New York: Freeman.

Deygers, B. (2019). Fairness and social justice in English language assessment. In X. Gao (Ed.), *Second handbook of English language teaching*. New York: Springer.

Fernandes, A., Anhalt, C. O., & Civil, M. (2009). Mathematical interviews to assess Latino students. *Teaching Children Mathematics, 16*(3), 162–169.

Franklin, C., et al. (2007). *Guidelines for assessment and instruction in statistics education (GAISE) report: A pre-k-12 curriculum framework*. Alexandria, VA: American Statistical Association.

GAISE College Report ASA Revision Committee. (2016). *Guidelines for assessment and instruction in statistics education college report 2016*. http://www.amstat.org/education/gaise.

Garaway, G. B. (1994). Language, culture, and attitude in mathematics and science learning: A review of literature. *Journal of Research and Development in Education, 27*(2), 102–109.

Gephart, R. P., Jr. (1988). *Ethnostatistics: Qualitative foundations for quantitative research*. Newbury Park, CA: Sage.

Groth, R. E. (2019). Applying design-based research findings to improve the Common Core State Standards for data and statistics in grades 4-6. *Journal of Statistics Education, 27*(1), 29–36.

Halvorsen, K. T., & Moore, T. L. (2005). Motivating, monitoring, and evaluating student projects. In T. L. Moore (Ed.), *Teaching statistics: Resources for undergraduate instructors* (pp. 27–34). Washington, DC: Mathematical Association of America.

Huberty, C. J. (2000). Judgment in quantitative research. *The Mathematics Educator, 10*(1), 5–10.
Irazábal, A. (2018). Stressing polysyllabic words: Un-academic activities for an academic matter. *Ideas, 4*(4), 1–11.
Joiner, B. L. (1988). Let's change how we teach statistics. *Chance, 1*(1), 53–54.
Kaplan, J. J., Fisher, D. G., & Rogness, N. T. (2009). Lexical ambiguity in statistics: What do students know about the words association, average, confidence, random and spread? *Journal of Statistics Education, 17*(3), 1–19.
Kaplan, J. J., Rogness, N. T., & Fisher, D. G. (2012). Lexical ambiguity: Making a case against spread. *Teaching Statistics, 34*(2), 56–60.
Lesser, L. (2007). Critical values and transforming data: Teaching statistics with social justice. *Journal of Statistics Education, 15*(1), 1–21.
Lesser, L. (2008). Social justice, gender equity, and service learning in statistics education: Lessons learned from the DOE-funded Project ACE (ACtion for Equity). In *Proceedings of the 2008 Joint Statistical Meetings, Section on Statistical Education* (pp. 424–431). Alexandria, VA: ASA. http://ww2.amstat.org/membersonly/proceedings/papers/300471.pdf.
Lesser, L. (2010a). Equity and the increasingly diverse tertiary student population: Challenges and opportunities in statistics education. In C. Reading (Ed.), *Proceedings of the Eighth International Conference on Teaching Statistics.* Voorburg, The Netherlands: International Statistical Institute.
Lesser, L. (2010b). The necessity of equity in teaching statistics. *Philosophy of Mathematics Education Journal, 25*, 1–8.
Lesser, L. (2010c). An ethnomathematics spin on statistics class. *Notices of the North American Study Group in Ethnomathematics, 3*(2), 5–6.
Lesser, L. (2011a). Supporting learners of varying levels of English proficiency. *Statistics Teacher Network, 77*, 2–5.
Lesser, L. (2011b). Low-tech, low-cost, high-gain, real-time assessment? It's all in the cards, easy as ABCD! *Texas Mathematics Teacher, 58*(2), 18–22.
Lesser, L. (2013a). Using context to classify variables. *Teaching Statistics, 35*(1), 29–31.
Lesser, L. (2013b). Dare to share: Tell us how you've used *TEEM* articles. *Teaching for Excellence and Equity in Mathematics, 5*(1), 35.
Lesser, L. (2014). Staring down stereotypes. *Mathematics Teacher, 107*(8), 568–571.
Lesser, L. (2015). Learning language: A mathematics educator's reflection on empathy and privilege. *Teaching for Excellence and Equity in Mathematics, 6*(1), 25–32.

Lesser, L. (2016). More statistics education literature on analogies. *The American Statistician, 70*(4), 434.
Lesser, L. (2018). Modulating misconceptions by musical means. *Teaching Statistics, 40*(3), 79–82.
Lesser, L., & Kephart, K. (2011). Setting the tone: A discursive case study of problem-based inquiry learning to start a graduate statistics course for in-service teachers. *Journal of Statistics Education, 19*(3), 1–29.
Lesser, L. M., Pearl, D. K., Weber, J. J., Dousa, D. M., Carey, R. P., & Haddad, S. A. (2019). Developing interactive educational songs for introductory statistics. *Journal of Statistics Education, 27*(3), 238–252. https://www.tandfonline.com/doi/full/10.1080/10691898.2019.1677533.
Lesser, L., & Wagler, A. (2016). Tools for assessing readability of statistics teaching materials. *Journal of Computers in Mathematics and Science Teaching, 35*(2), 153–171. Also, see https://www.CAUSEweb.org/cause/ecots/ecots16/posters/b/5.
Lesser, L., Wagler, A., & Abormegah, P. (2014). Finding a happy median: Another balance representation for measures of center. *Journal of Statistics Education, 22*(3), 1–27.
Lesser, L., Wagler, A., Esquinca, A., & Valenzuela, M. G. (2013). Survey of native English speakers and Spanish-speaking English language learners in tertiary introductory statistics. *Statistics Education Research Journal, 12*(2), 6–31.
Lesser, L., Wagler, A., & Salazar, B. (2016). Flipping between languages? An exploratory analysis of the usage by Spanish-speaking English language learner tertiary students of a bilingual probability applet. *Statistics Education Research Journal, 15*(2), 145–168.
Lesser, L., & Winsor, M. (2009). English language learners in introductory statistics: Lessons learned from an exploratory case study of two pre-service teachers. *Statistics Education Research Journal, 8*(2), 5–32.
Loux, T., & Gibson, A. K. (2019). Using Flint, Michigan lead data in introductory statistics. *Teaching Statistics, 41*(3), 85–88.
Mocko, M., Lesser, L., Wagler, A., & Francis, W. (2017). Assessing effectiveness of mnemonics for tertiary students in a hybrid introductory statistics course. *Journal of Statistics Education, 25*(1), 2–11.
Moore, D. S. (1988). Should mathematicians teach statistics? *The College Mathematics Journal, 19*(1), 3–7.
National Council of Teachers of Mathematics. (2000). *Principles and standards for school mathematics*. Reston, VA: NCTM.

Nolan, V. (2002). Influence of attitude towards statistics, English language ability and mathematical ability in the subject quantitative techniques at the Vaal Triangle Technikon, South Africa. In B. Phillips (Ed.), *Proceedings of the Sixth International Conference on Teaching Statistics: Developing a statistically literate society*, Cape Town, South Africa [CD-ROM]. Voorburg, The Netherlands: International Statistical Institute.

Ortiz, J. J., Cañizares, M. J., Batanero, C., & Serrano, L. (2002). An experimental study of probabilistic language in secondary school textbooks. In B. Phillips (Ed.), *Proceedings of the Sixth International Conference on Teaching Statistics: Developing a statistically literate society*, Cape Town, South Africa [CD-ROM]. Voorburg, The Netherlands: International Statistical Institute.

Pfannkuch, M., Regan, M., Wild, C., & Horton, N. J. (2010). Telling data stories: Essential dialogues for comparative reasoning. *Journal of Statistics Education, 18*(1), 1–38.

Rangecroft, M. (2002). The language of statistics. *Teaching Statistics, 24*(2), 34–37.

Roberts, H. V. (1992). Student-conducted projects in introductory statistics courses. In F. and S. Gordon (Eds.), *Statistics for the twenty-first century* (pp. 109–121). Washington, DC: Mathematical Association of America.

Rubel, L., Driskill, M., & Lesser, L. (2012). Decennial redistricting: Rich mathematics in context. *Mathematics Teacher, 106*(3), 206–211.

Rumsey, D. J. (2009). Watching our language when we teach statistics. *Journal of Statistics Education, 17*(1), 1–3.

Short, T. H., & Pigeon, J. G. (1998). Protocols and pilot studies: Taking data collection projects seriously. *Journal of Statistics Education, 6*(1).

Smith, D. J. (2002). *If the world were a village: A book about the world's people.* Tonawanda, NY: Kids Can Press.

Sullivan, M. (2010). *Statistics: Informed decisions using data* (3rd ed.). Boston: Pearson.

Viera, J. (2019). Emergent bilinguals' engagement in an online mathematics course utilizing an intelligent tutoring system. Unpublished doctoral dissertation, The University of Texas at El Paso, El Paso, Texas.

Wagler, A. E., & Lesser, L. M. (2011). Teaching statistics to culturally and linguistically diverse students. In *Proceedings of the 2011 Joint Statistical Meetings, Section on Statistical Education* (pp. 821–830). http://www.statlit.org/pdf/2011Wagler-Lesser-JSM.pdf.

Willis, S. (1995). Science education: Reformers promote hands-on, inquiry-based learning. *Curriculum Update, 14*(3), 1–8.

4

Designing Professional Development to Support Teachers' Productive Beliefs, Knowledge, and Practices for Teaching Mathematics to English Language Learners

Maria I. Ruiz and Melissa A. Gallagher

Supporting all students to develop high levels of mathematics proficiency should be the goal of every mathematics educator; however, not all teachers have the beliefs and knowledge to choose instructional practices that reach all learners. English language learners (ELLs) may encounter teachers who do not understand their cultural and linguistic backgrounds and specific educational needs (de Araujo, 2017). These challenges can be compounded when teachers do not believe that all students can achieve at high levels (Archambault, Janosz, & Chouinard, 2012). As the population of students who speak a language other than English at home rises in US schools (U.S. Department of Education, 2016), professional development (PD) designers must create experiences that provide teachers with knowledge for teaching mathematics, help teachers

M. I. Ruiz (✉)
University of Louisiana at Lafayette, Lafayette, LA, USA
e-mail: maria.ruiz@louisiana.edu

M. A. Gallagher
University of Houston, Houston, TX, USA
e-mail: magallag@central.uh.edu

L. C. de Oliveira and M. Civil (eds.), *Teaching Mathematics to English Language Learners*, https://doi.org/10.1007/978-3-030-48355-5_4

develop productive beliefs about the nature of mathematics as a process of inquiry, and support teachers' experimentation with instructional practices and strategies that enhance ELLs' mathematics understanding.

In the state where the PD initiative described in this chapter took place, the percentage of ELLs attending K-12 schools is relatively small (under 5%) compared to the national average. However, the influx of immigrant families in specific regions of the state has resulted in a disproportionate increase in the number of ELLs being served by some schools. This rapid change in student demographics is often a cause of concern among teachers and school administrators in those areas due to their limited resources, lack of training in working with this population, and limited exposure to individuals and families from other cultures. Local teachers' anecdotal reports of the practices they use teaching ELLs suggest a narrow understanding of ELLs' instructional needs and of recommended teaching strategies for facilitating ELLs' learning of academic content. Therefore, a need was identified for PD that focused on increasing local teachers' knowledge and skills in the teaching of mathematics and providing explicit instruction of recommended teaching strategies to address ELLs' learning needs. This chapter describes the development and implementation of the Supporting English Learners' Achievement in Mathematics (SELAM) PD initiative, which sought to support the development of beliefs, knowledge, and practices to promote ELLs' learning of mathematics.

Teachers' Productive Beliefs, Knowledge, and Practices

Teachers must have productive beliefs, knowledge, and practices to support the learning of all students. Some productive beliefs that teachers of mathematics need involve a growth mindset (Dweck, 2008), including the expectation that their students can achieve (Archambault et al., 2012), and a belief in the nature of mathematics as a process of inquiry, rather than a set of procedures to be memorized (Tatto et al., 2012). However, supporting ELLs requires additional beliefs, including valuing linguistic diversity (Lucas & Grinberg, 2008) and a commitment to advocating for the needs of ELLs (Athanases & de Oliveira, 2007). Many teachers of mathematics need support in developing these beliefs.

In addition to productive beliefs, teachers of mathematics need specific knowledge to support the learning of all students effectively. When working with ELLs, teachers should have knowledge of their content area (Viesca, Joseph, & Commins, 2019) and knowledge of second language learning (Lucas, 2011). The specific knowledge teachers need is referred to as mathematical knowledge for teaching (MKT; Ball, Thames, & Phelps, 2008), which includes an understanding of the connectedness across mathematical domains and grade levels and how to help students see these connections. Besides having strong MKT, mathematics teachers of ELLs must understand the process of second language learning (Lucas, 2011), including the stages ELLs will navigate (Krashen & Terrell, 1983), and how they can support students as they move through those stages. Knowledge about this developmental process can help teachers support ELLs in the mathematics classroom.

When teaching mathematics to ELLs, there are practices teachers can implement which, through the frame of productive beliefs (Fives & Buehl, 2012) and strong MKT, can help them to support students effectively. For instance, using multimodal communication (e.g., gestures, visuals, multiple representations), while focusing on academic language development, can support ELLs' learning of mathematics (Miller & Warren, 2014). Modeling mathematical thinking can help ELLs develop and access mathematics discourse (Hansen-Thomas, 2009; Khisty & Chaval, 2002). Additionally, using ELLs' primary language can promote understanding (de Araujo, Roberts, Willey, & Zahner, 2018). Ramirez and Celedón-Pattichis (2012) suggested guiding principles for supporting mathematical proficiency for ELLs: (a) implementing challenging mathematical tasks; (b) fostering linguistically sensitive social environments; (c) supporting learning English while learning mathematics; (d) using mathematical tools and modeling as resources; and (e) upholding cultural and linguistic differences as intellectual resources. Also, Driscoll, Nikula, and DePiper (2016) reiterated the value of incorporating visual representations, language access strategies, and language production strategies into mathematics instruction to support ELLs' achievement in mathematics. The SELAM PD focused on both general

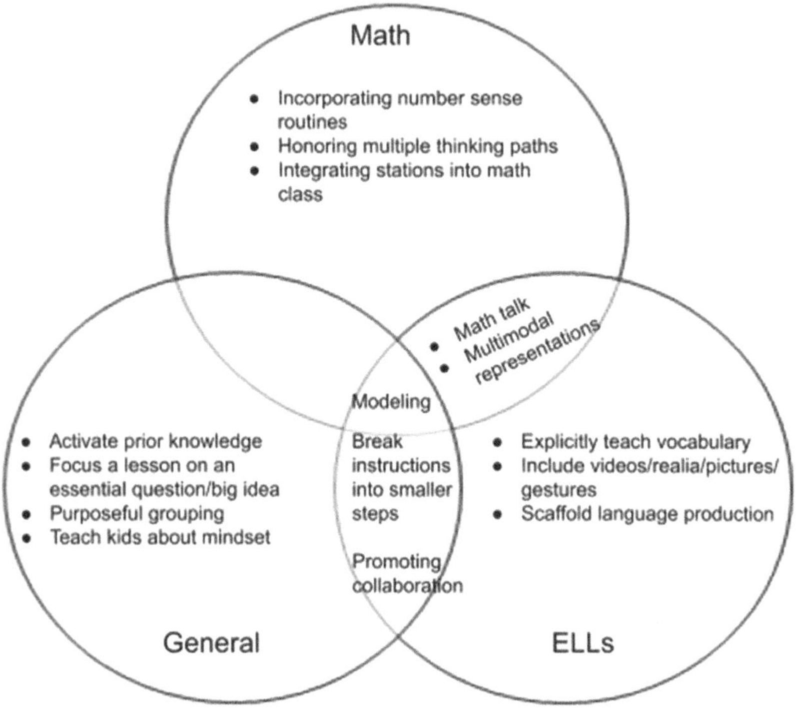

Fig. 4.1 Mathematics-specific, general, and ELL-specific instructional strategies covered in SELAM

instructional practices for ELLs and mathematics-specific instructional practices (Fig. 4.1).

Professional Development

Teacher knowledge, beliefs, and instructional practices can be affected through PD (Gomez Zwiep & Benken, 2013; Polly et al., 2015). Targeted PD can increase teachers' cultural awareness and knowledge, potentially leading them to engage in culturally responsive teaching (Parker, Bartell, & Novak, 2017). Philipp (2007) noted two impediments to changing teachers' beliefs:

> [C]oncern for the well-being of children that often inhibits teachers' willingness to challenge students and difficulty in overcoming the classroom challenges that derive from moving beyond their role as the teacher as one whose job it is to tell students how to be successful. (p. 281)

These two impediments are rooted in teachers' desire to be caring and not make students uncomfortable as they productively struggle with new ideas, as well as some teachers' orientation toward their role as the expert who is responsible for all learning happening in the classroom. To affect change in teachers' practices, PD designers must address teachers' beliefs, providing opportunities for them to enact practices successfully and to reflect on that enactment (Fives & Buehl, 2012; Philipp, 2007).

Effective PD focuses on the specific content teachers will be teaching and how students learn that content, aligns with teachers' actual experiences in the classroom, uses curriculum materials and assessments, provides opportunities to observe and analyze students' understanding of the subject matter, and is spread over time (Garet, Porter, Desimone, Birman, & Yoon, 2001). Loucks-Horsley, Stiles, Mundry, Love, and Hewson (2009) suggest that before conducting PD in mathematics, PD designers must examine how their own beliefs and knowledge on the subject inform the PD they will be leading. Designers must evaluate the context where the PD will be offered, determine teachers' needs, and identify critical issues that must be addressed. Loucks-Horsley and colleagues emphasize that designers should then use this information to select strategies that align with the goals and vision for the PD. These recommendations informed the process of designing and implementing SELAM PD activities.

The SELAM Initiative

After discussing with district leaders the need for specific PD that addressed mathematics instruction for ELLs, we outlined a structure for SELAM: (a) a summer institute at the beginning of SELAM followed by another one during the summer of year 2 of the project, embedded within an existing summer school program for ELLs; (b) three instructional coaching visits throughout the school year; and (c) three workshops during the school year (Fig. 4.2). During the first summer institute, we developed participants' foundational knowledge and skills for teaching mathematics to ELLs in order to facilitate their participation in the PD activities that would follow during the school year. During instructional coaching visits, we used the transformative coaching model (Aguilar, 2013) rather than more traditional classroom observations to encourage teachers to take ownership of their professional learning. The second summer institute combined presentations, workshops, coaching training, and a mathematics summer camp for ELLs. The summer camp provided participants with repeated opportunities to apply their new knowledge about ELLs and mathematics instruction and to practice the skills developed during the first year of the PD. Also, during the second summer institute, participants were trained to serve as coaches for their colleagues, in order to build capacity within the district.

After presenting the PD outline to district administrators, we worked with the Supervisor of Specially Funded Programs and the Title I Specialist who coordinated the English as a second language (ESL) program for the district to plan the SELAM initiative around the district's needs. We agreed that SELAM would have a threefold purpose:

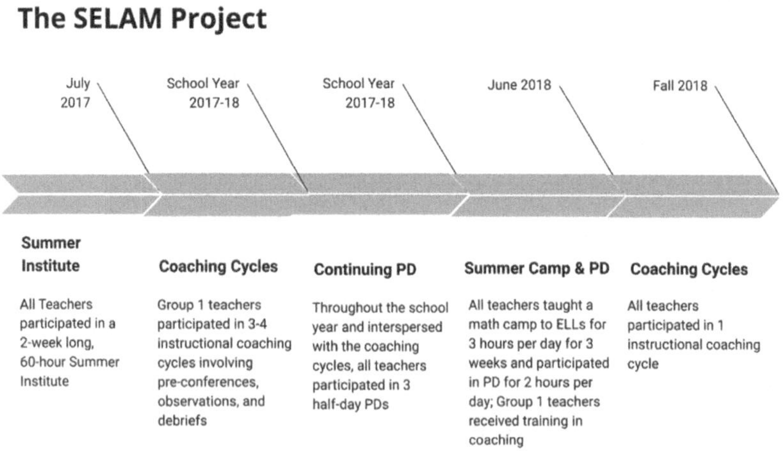

Fig. 4.2 Overview of SELAM PD initiative

(a) to increase teachers' MKT for rational numbers, (b) to increase teachers' understanding of cultural and linguistic factors that affect ELLs' ability to succeed in the mathematics classroom, and (c) to develop teachers' knowledge of instructional practices that facilitate the learning of mathematics for ELLs. Rational numbers, especially fractions, is a challenging area for most elementary teachers (Ball, 1990; Borko et al., 1992; Van Steenbrugge, Lesage, Valcke, & Desoete, 2014) and an area of weakness for American students (Izsák, Tillema, & Tunç-Pekkan, 2008; Lamon, 1993), including ELLs in the partnering district. Thus, the decision was made to focus on developing the rational numbers content and pedagogical content knowledge of participating teachers. Based on the growing understanding of the setting where the project would take place and our knowledge of effective elements of PD in education, we selected four principles to guide the design of SELAM:

Principle 1: Select participants purposefully. Although we believe that all teachers deserve to participate in PD that supports their productive beliefs, knowledge, and practices for working with ELLs in mathematics, when resources are limited, participants should be selected who are most open to learning about new ways to improve their practices and, thus, most likely to benefit from the PD experiences.

Principle 2: Embed activities/information on productive beliefs, knowledge, and practices throughout each institute/workshop day, revisiting topics as necessary. Embedding activities that address beliefs, knowledge, and practices throughout each day allows PD designers to spiral ideas as well as to foster incremental changes in knowledge, beliefs, and practices over time.

Principle 3: Create opportunities for participants to experience instructional practices as learners. One impediment to changing teachers' beliefs is their desire to make sure students are comfortable and, therefore, not create adequate challenges for them (Philipp, 2007). By allowing teachers to participate in learning experiences as learners, they can reflect on their own emotional engagement during those experiences, recognizing productive struggle as integral to learning (National Council of Teachers of Mathematics [NCTM], 2014).

Principle 4: Create opportunities for participants to enact and reflect on the instructional practices, supported by a coach, in both traditional and non-traditional educational settings. Reflection is necessary for changing teachers' beliefs and practices (Fives & Buehl, 2012; Philipp, 2007). Coaching is a structure that can foster teacher reflection (Aguilar, 2013). PD that includes individual coaching has been shown to have a greater impact on teacher practice than participating in PD without coaching (Neuman & Cunningham, 2009).

Although we considered many of the other principles and guidelines for creating PD described in the literature (Garet et al., 2001; Sztajn, Borko, & Smith, 2017), these principles were central to the design and implementation of all activities included in the SELAM initiative.

The SELAM PD Program

All activities completed during the SELAM program were guided by the principles described above.

Selection of participants [Principle 1]. We introduced the project to principals from the four schools with the largest ELL populations in the district at a meeting arranged by the district partners. Principals were asked to recommend and encourage mathematics teachers in grades 3–8 at their schools, who had at least one ELL in their classes, to join the project. At the end of the 2017 summer institute, participants were given the choice to participate in the rest of the program as part of one of two groups. Participants in Group 1 were coached throughout the school year, attended the 2018 summer institute for three weeks, and were trained as coaches. Group 2 teachers did not receive any coaching during the school year; they attended weeks 2 and 3 of the 2018 summer institute and were coached by Group 1 teachers. All 16 participants attended the SELAM workshops offered throughout the school year. Ten participants self-selected to be part of Group 1 and six participants chose to be in Group 2. Based on comments shared by participants at the end of both summer institutes, the main motivation for selecting to be part of Group 1 was participants' interest in learning more about coaching.

The 2017 summer institute [Principles 2 and 3]. The first SELAM PD activity was a summer institute with face-to-face meetings for six hours a day for eight days, and 12 hours of online modules. Each day, participants took part in activities designed to increase their understanding of (a) the research supporting growth mindset; (b) MKT related to the instruction of rational numbers; (c) the cultural and linguistic factors that can influence ELLs' learning and academic performance, including the second language acquisition process; (d) recommended strategies for teaching mathematics to ELLs; and (e) modifying their curricular materials to integrate these strategies. This PD was responsive (Anderson & Gallagher, 2019) in that activities were modified each day to meet participants' emerging needs.

Productive beliefs: Growth mindset. To address participants' beliefs about mathematics learning and teaching, the designers introduced participants to the concept of mindset. The PD included regular discussions about: (a) approaching mathematics instruction with a growth mindset (believing that learning and ability improve through effort and work), as opposed to a fixed mindset (believing that abilities and talents are fixed traits); (b) the connections between mindset, beliefs, and student performance in mathematics (Boaler, 2015); and (c) the role of neuroplasticity in learning. Participants engaged in daily activities designed to help them identify views and behaviors indicating a fixed mindset about their own and students' mathematics abilities and to provide practice in using strategies that promote a growth mindset in the mathematics classroom. Other activities included examining case scenarios for examples of mindset, self-assessing to identify potential areas for mindset growth which might improve their outcomes when teaching mathematics to all students, including ELLs, and reflecting daily about new mindset information they learned. Also, participants completed an online module (Stanford University, 2014), which focused on developing a growth mindset in mathematics.

Understanding ELLs' needs. A section of the PD was dedicated to increasing participants' understanding of factors that can affect ELLs' academic performance. It included an exploration of participants' background knowledge about ELLs' characteristics and instruction, followed by a facilitated discussion to confirm or disconfirm their preconceptions. Participants learned about factors that influence ELLs' learning and performance in school, second language acquisition, steps for differentiating between limited proficiency in English and a language disorder, guidelines for selecting supports and providing quality instruction to ELLs, selecting appropriate accommodations, and differentiating between accommodations and instructional practices used with ELLs. In addition, designers described and modeled practices that participants could incorporate into their teaching to facilitate ELLs' learning and participation in mathematics class activities (Fig. 4.1). Participants also completed an online module (Vanderbilt University, 2017) which provided information on recommended practices to use with ELLs.

Knowledge of mathematics content. During the mathematics content knowledge section of the PD, the mathematics workshop model (Lempp, 2017) was used to immerse participants in experiencing the mathematical content as learners. The use of mathematical tasks and engaging in those tasks as learners and then reflecting on them is a powerful tool in teacher learning (Sztajn et al., 2017). Each day, one of the designers began the session with a number sense routine, then reviewed prerequisite concepts and explicitly taught vocabulary related to the lesson. Next, the designer presented the group with an essential question and explicitly taught selected content related to rational numbers. Then, participants broke into small groups and rotated through stations. At the stations, they used manipulatives to solve problems, worked on rich mathematics tasks, and engaged in mathematics talk with peers. Activities and stations included in the mathematics workshop routinely incorporated and modeled strategies of ELL instruction discussed during the morning sessions (e.g., using visuals and manipulatives, providing sentence frames to support communication about content, simplifying language). See Fig. 4.3 for a sample task used in a session focused on proportional reasoning. This task integrated the following instructional

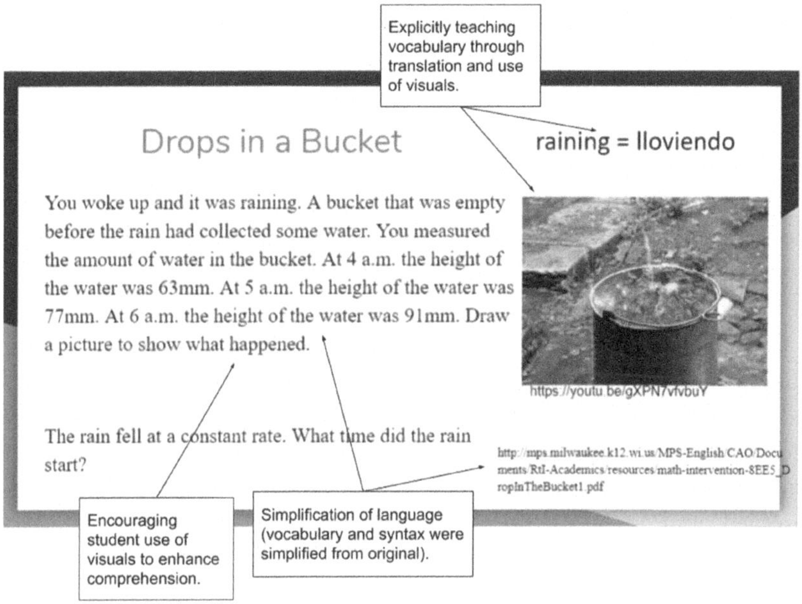

Fig. 4.3 Proportional reasoning task with ELL instructional practices identified

strategies: using visuals, explicitly teaching vocabulary, and simplifying language without reducing the cognitive demand of a task.

At the end of each session, the designer encouraged participants to discuss aspects of the mathematics content learned and the activities completed. The designer helped participants focus on the main ideas related to the content and strategies used, as well as elements of the lesson where she had integrated instructional strategies to support ELLs. The content knowledge sections supported participants in developing deeper content knowledge, unearthed discussions about common student misconceptions, and modeled the integration of effective ELL and mathematics instructional practices.

School year: Coaching cycles and workshops [Principles 2, 3, and 4]. During the 2017–2018 school year, the designers conducted at least three rounds of instructional coaching cycles with each of the Group 1 participants. Between each round of coaching cycles, the designers held half-day workshops with Group 1 and Group 2 participants, which included hands-on activities to impact participants' productive beliefs, knowledge, and practices. Integrating coaching cycles and workshops provided opportunities to tailor the workshops to needs observed during the coaching cycles.

Coaching cycles. The designers used the transformative coaching model (Aguilar, 2013), which situates the teacher as knowledgeable and positions the coach as facilitating the teacher to leverage her own knowledge to meet her goals. Each coaching cycle included: (a) a pre-conference in which the participant set goals, discussed the lesson to be observed, and asked the coach to focus on a specific teaching practice; (b) an observation during which the coach took detailed observational notes, focusing on the participant's personal goals; and (c) a debrief conference in which the participant reflected on her instruction and the coach shared data from the observation.

Workshops. The first workshop was held one month after the start of the 2017–2018 school year. It maintained the same format used during the summer institute with separate blocks of time dedicated to growth mindset, ELL instruction, effectively using mathematics stations, increasing participants' MKT, and reflection. To help participants apply their knowledge of mindset to their work with their new students, the designers had participants review their class rosters and select any student they found to be most challenging to teach. Then the participants were asked to answer several questions about the focus student. The questions helped participants reflect on the student's positive qualities and encouraged them to approach the student's challenges with a growth mindset. After participants shared their experiences with the activity, the designer provided a short review of mindset concepts bringing attention to the need for regular self-assessment to avoid reverting to beliefs that might hinder the quality of instruction provided to students. Next, the designers reviewed the steps involved in modifying lessons to meet the learning and linguistic needs of ELLs, as well as some strategies they could use. For the mathematics content session, one designer provided detailed guidelines for creating and integrating stations in mathematics classes. This section was added to the workshop in response to requests from some participants who had shown interest in implementing this practice in their classrooms. The mathematics content section of the workshop focused on the division of fractions, which most participants expressed being a weakness for them when it was initially taught at the summer institute.

During Workshop 2, mindset and ELL instructional strategies were addressed through a group case study activity. Each group was assigned a detailed case study that described a hypothetical teacher's thinking process while designing and implementing mathematics lessons in a classroom with ELLs. Participants were asked to analyze the adequacy of the teacher's selected practices, identify instances when the teacher had shown or promoted a growth or fixed mindset about mathematics learning or students, and provide suggestions for improving the lesson. For the content knowledge portion of this workshop, participants solved rich mathematics tasks on proportional reasoning (e.g., Fig. 4.3) and writing linear equations. Workshop 2 ended with a demonstration lesson in a sixth-grade class at one of the participating schools. A designer taught the lesson, which focused on the adaptation of a lesson in the district-mandated curriculum that integrated instructional practices discussed in SELAM. After the lesson, participants debriefed observed practices with guidance from the designers.

For the third workshop, designers guided participants through the steps for writing language goals for their mathematics lessons. Although this skill had been the focus of previous activities, the designers reviewed this concept after observing that participants continued to struggle to identify and explicitly address the language requirements of lessons. This activity was followed by a presentation on structures and implementation of mathematics workshop. After the presentation, the designer modeled a mathematics workshop while reviewing content on equivalent fractions, operations with fractions, and proportional reasoning. The last hour of the workshop was spent practicing how to select rich mathematics stations assisted by an assessment tool developed by the designers. To facilitate participants' implementation of stations in their classrooms, the SELAM team created and gave each participant a ready-to-go kit with stations appropriate for the grade level they taught. The workshops provided an opportunity for the designers to address instructional needs that they noticed, or participants identified, during the coaching cycles.

The 2018 summer institute [Principles 2, 3, and 4]. The second SELAM summer institute presented an opportunity for participants to work with students in a non-traditional educational setting. For this institute, the designers collaborated with the school district to add a mathematics component to the academic (reading and writing) summer camp they regularly offered for ELLs. The 2018 summer institute ran six hours per day for three weeks. The mathematics camp, which was embedded within the institute, ran for three hours on every day of the summer institute; it was open to all ELLs attending grades 2–12 in the district. The camp offered participating teachers the opportunity to implement the knowledge and skills they had learned throughout SELAM and to receive feedback from the designers without the pressures associated with all the other tasks they were expected to complete during the school year. Group 1 participants attended weeks 1–3, whereas Group 2 participants attended the camp portion of the institute during week 2 and the morning PD and mathematics camp during week 3, but did not participate in week 1.

Morning PD. For the full length of the second summer institute, the first two hours of each day were dedicated to engaging participants in activities designed to (a) increase their MKT of rational numbers on topics not covered during the first institute or workshops, (b) reinforce their knowledge of teaching practices to support ELLs, (c) strengthen their understanding of the importance of maintaining and promoting a growth mindset when teaching mathematics to ELLs, and (d) introduce them to co-teaching models. During weeks 1 and 2, daily activities were included in the morning routine to prepare Group 1 participants to function as coaches for their peers. Likewise, throughout week 3, an increased emphasis was placed on helping participants integrate the knowledge and skills developed during their participation in the SELAM project into lesson planning and daily instruction.

Daily PD also addressed areas of need identified during classroom observations and interaction with participants, including gaps or misconceptions participants seemed to have regarding specific content on rational numbers; ELL teaching strategies to use with particular students; effective co-teaching strategies to use during specific activities participants had integrated into their lessons; differentiated planning and instruction; cultural aspects of student social and learning behavior; and reminders to maintain a growth mindset with students who were underperforming, had very limited English language skills, or exhibited socially challenging behaviors. Participants were also given time each day to plan.

Mathematics camp for ELLs. The week before the second summer institute, Group 1 participants were assigned to co-teaching pairs based on the grade levels they taught during the school year. They were given one day to review instructional practices, co-plan lessons, and prepare materials for the mathematics summer camp. Although participants had the freedom to create lessons on any mathematical concepts they deemed appropriate for their assigned grade level and individual students, they were asked to include an outdoor mathematics activity, a rich mathematics task, and a mathematics workshop with stations in the daily schedule. For the next three weeks, teachers in Groups 1 and 2 completed various activities and shared teaching responsibilities associated with the SELAM mathematics summer camp (Appendix 1).

Developing coaching skills [Principle 4]. Group 1 participants worked on developing coaching skills during the entire 2018 summer institute. They participated in a variety of activities designed to develop their knowledge of the transformative coaching model, increase their understanding of the benefits of coaching, and provide enough practice in their implementation of coaching techniques.

Week 1. Group 1 participants took turns presenting chapters from *The Art of Coaching* (Aguilar, 2013) during the morning PD throughout the week. On days 2 and 3 of the mathematics camp, Group 1 participants in each co-teaching team took turns teaching the class and observing and coaching her co-teacher. Group 1 participants acting as coaches were accompanied by one of the PD designers for their first coaching cycle, which created an opportunity to coach the new coaches. On days 4 and

5, participants took turns teaching the class and observing and coaching participants in other co-teaching teams.

Week 2. The designers led follow-up activities to strengthen Group 1 participants' knowledge of instructional coaching strategies and their coaching skills. Group 1 participants continued to take turns presenting chapters from *The Art of Coaching* (Aguilar, 2013) during the morning PD and coaching each other outside their co-teaching teams.

Week 3. Group 1 participants continued to coach other Group 1 participants that were not part of their co-teaching team. They also took turns with their Group 1 partners to coach the Group 2 participants assigned to their teams.

Lessons Learned

Overall, we felt that SELAM was a very successful project. We observed numerous instances of participants implementing better practices for supporting ELLs in mathematics. At least two participants chose to pursue additional teaching certification in English as a Second Language and to enter master's programs after completing SELAM. In this section, we share some of our takeaways and impressions of the project, especially as they relate to Principles 1–4. We hope these lessons learned will help other PD designers to create PD to support teachers' productive beliefs, knowledge, and practices for teaching mathematics to ELLs. Empirical research on changes to teachers' beliefs, knowledge, and practices are reported elsewhere (Gallagher & Ruiz, under review).

Our goal with Principle 1 was to select participants who would most benefit from this PD experience. We asked principals from the schools to nominate teachers to participate. We had hoped to spend our finite resources on teachers who would be most likely to take up the practices. Unfortunately, we did not clearly communicate that we wanted to start this work with teachers who would be most open to change. We later discovered that some of the teachers had not independently chosen to participate, but had rather done so upon insistence from their principals. We learned that others had been told that they would be telling us about their teaching practices rather than learning about practices to support ELLs. Given that PD that is aligned with teachers' personal goals is more effective (Garet et al., 2001), we believe this misalignment may have hindered some of the teachers' ability to shift their practices. In the future, it would be important to communicate with teachers directly during the recruiting phase and ensure that teachers' participation in the program is guided by their own desire to change their practices. Additionally, although we communicated with principals about the goals of SELAM and developed these goals in conjunction with the district, we later found out that some principals were telling participants not to differentiate or use stations to teach in small groups. We did not do enough to involve multiple district leaders and administrators throughout the project. We would revise Principle 1 to be: When faced with finite resources, work with administrators to choose participants whose personal and professional goals align with those of the PD. We would also add Principle 5: Involve administrators and district leaders throughout the PD, ensuring they understand the goals of the project and the rationale for those goals.

Principles 2, 3, and 4 were designed to support teacher change through responsive PD. Because of these principles and our belief in the need to be responsive to teachers' needs, we shifted our PD plans at the end of each day. We felt that the most powerful part of SELAM was that we were able to shift to meet participants' needs as they emerged and that this responsiveness may have led to changes in participants' practices. For instance, one participant was already differentiating instruction and using mathematics work stations before SELAM began. However, her stations were focused on procedural fluency and did not include scaffolds for ELLs. By the end of SELAM, her stations varied so that some focused on procedural fluency, whereas others focused on building conceptual understanding. Most stations included visual as well as linguistic scaffolds. She felt very proud of the effort she had made to come up with new stations to better meet her students' needs. We advocate for responsive PD for teachers designed using Principles 2, 3, and 4.

Principle 2 addresses the idea of PD focused on beliefs, knowledge, and practices. We were able to openly discuss the impact of participants' mindset about the content they taught, their students, and themselves as teachers of mathematics without making participants feel guilty or judged because of our non-evaluative role (i.e., being outside the district). It was great seeing the light bulbs go off every time a participant realized her biases, identified mistakes she had been making, or showed interest in learning more about a topic we were discussing (e.g., brain and learning, mindset, how cultural and linguistic factors could affect ELLs' learning needs). As the concept of growth mindset became an intrinsic component of all the work done during SELAM activities, participants seemed to be more comfortable acknowledging their own struggles and more open to receiving feedback from designers and peers. Embedding all three components that impact practice—knowledge, beliefs, and the practices themselves—within each day of the PD facilitated opportunities for incremental change.

One of the most powerful aspects of SELAM was the mathematics camp for ELLs. We created this aspect of SELAM with Principle 4 in mind. What we had not anticipated was just how readily participants would implement the practices when co-teaching in a non-traditional setting. Although the participants had participated in SELAM for 11 months at the time of the mathematics camp, we observed very few practices being implemented during our coaching visits. However, during the mathematics camp, we saw all participants using small group instruction, integrating visuals, honoring students' home languages, using realia, and so much more. We believe that giving the teachers an opportunity to co-teach and co-plan, sharing resources with them, and giving them a low-stakes environment in which to implement these practices created the opportunities necessary for participants to try out the practices and see their value. We are now ardent proponents of creating low-stakes environments in which teachers can try out the practices they are learning.

Lastly, if there were one thing we would change, it would have been the duration of the project. After 14.5 months, many participants were just beginning to shift their practices. Had we been able to continue to work with these teachers, then we possibly could have facilitated more shifts in practice. We would have liked to have used the Group 1 participants to coach Group 2 participants throughout the second year of the project and worked with the district to support these teachers in taking on leadership roles within the district. For instance, the SELAM participants could have worked together to plan PDs for the teachers at their respective schools on the beliefs, knowledge, and practices they learned about through SELAM. We would add Principle 6: PD should be sustained, creating increasing opportunities for leadership for teachers over time.

Based on our experience with SELAM, we offer these six principles for designing PD to support teachers of mathematics' development of productive beliefs, knowledge, and instructional practices necessary to effectively engage with ELLs:

Principle 1: When faced with finite resources, work with administrators to choose participants whose personal professional goals align with those of the PD.

Principle 2: Embed activities/information on productive beliefs, knowledge, and practices throughout each institute/workshop day, revisiting topics as necessary.

Principle 3: Create opportunities for participants to experience instructional practices as students.

Principle 4: Create opportunities for participants to enact and reflect on the instructional practices, supported by a coach, in both traditional and non-traditional educational settings.

Principle 5: Involve administrators and district leaders throughout the PD, ensuring they understand the goals of the project and the rationale for those goals.

Principle 6: PD should be sustained, creating increasing opportunities for leadership for teachers over time.

Conclusions

We believe that, as suggested in the literature on PD for teachers of mathematics (American Educational Research Association [AERA], 2005), aligning PD activities with experiences that teachers would actually have in the classroom and providing opportunities for teachers to see students' responses and learning while engaging in the proposed practices (e.g., rich tasks) contributed to the effectiveness of SELAM in impacting teachers' productive beliefs, knowledge, and practices for supporting ELLs in mathematics. Furthermore, we believe that keeping the six principles in mind when designing PD can create more impactful experiences for teachers.

As the ELL population grows in the USA, teachers need strategies to support students' mathematics learning and their language development. Teacher educators and PD designers must support teachers in developing the knowledge, beliefs, and practices needed to help these students succeed in mathematics. PD experiences that integrate workshops, coaching, and informal learning experiences (e.g., a summer camp) may be key to helping teachers develop the productive beliefs, knowledge, and practices necessary to support ELLs in the mathematics classroom.

Acknowledgments This work was made possible by a grant from the Louisiana Systemic Initiatives Program (LaSIP Contract # 17-207-ULL-M). This chapter does not necessarily reflect the views or policies of the Louisiana Board of Regents. We thank the teachers, students, district partners, and project team members who contributed to this project.

Appendix 1

2018 SELAM Summer Institute
Mathematics Camp Teaching Schedule

	Day 1	Day 2	Day 3	Day 4	Day 5
Week 1					
Group 1	Co-teach mathematics to assigned class at the mathematics camp for ELLs				
	Participate in group and individual reflection on the day's activities after camp dismissal				
Group 2	-No participation-				
Week 2					
Group 1			Train Group 2 teachers on how to implement recommended mathematics instructional strategies with ELLs		
Group 2	Participate in small-group in-service session lead by the PD designers to prepare for the summer camp	Continue instructional rounds (City, Elmore, Fiarman, & Teitel, 2009) with a designer	Observe Group 1 teachers in classes closest to the grade level typically taught		Co-teach with Group 1 teachers in assigned class
	Participate in instructional rounds (City et al., 2009) with a designer		Co-plan next day's lessons with Group 1 teachers		
Week 3					
Group 1	Co-plan and co-teach lessons in assigned classes throughout the day				
Group 2					

References

Aguilar, E. (2013). *The art of coaching: Effective strategies for school transformation*. San Francisco, CA: Wiley.

American Educational Research Association (AERA). (2005). Teaching teachers: Professional development to improve student achievement. *Research Points, 3*(1).

Anderson, B., & Gallagher, M. A. (2019). Responsive professional development for vocabulary instruction. *Action in Teacher Education, 41*(4), 361–378. https://doi.org/10.1080/01626620.2019.1616632.

Archambault, I., Janosz, M., & Chouinard, R. (2012). Teacher beliefs as predictors of adolescents' cognitive engagement and achievement in mathematics. *The Journal of Educational Research, 105*(5), 319–328.

Athanases, S. Z., & de Oliveira, L. C. (2007). Conviction, confrontation, and risk in new teachers' advocating for equity. *Teaching Education, 18*(2), 123–136.

Ball, D. L. (1990). Prospective elementary and secondary teachers' understanding of division. *Journal for Research in Mathematics Education, 21*, 132–144. https://doi.org/10.2307/749140.

Ball, D. L., Thames, M. H., & Phelps, G. (2008). Content knowledge for teaching what makes it special? *Journal of Teacher Education, 59*(5), 389–407. https://doi.org/10.1177/0022487108324554.

Boaler, J. (2015). *Mathematical mindsets: Unleashing students' potential through creative math, inspiring messages, and innovative teaching*. Hoboken, NJ: Wiley.

Borko, H., Eisenhart, M., Brown, C. A., Underhill, R. G., Jones, D., & Agard, P. C. (1992). Learning to teach hard mathematics: Do novice teachers and their instructors give up too easily. *Journal for Research in Mathematics Education, 23*, 194–222.

City, E. A., Elmore, R. F., Fiarman, S. E., & Teitel, L. (2009). *Instructional rounds in education: A network approach to improving teaching and learning*. Cambridge, MA: Harvard Education Press.

de Araujo, Z. (2017). Connections between secondary mathematics teachers' beliefs and their selection of tasks for English language learners. *Curriculum Inquiry, 47*(4), 363–389.

de Araujo, Z., Roberts, S. A., Willey, C., & Zahner, W. (2018). English Learners in K–12 Mathematics education: A review of the literature. *Review of Educational Research, 88*(6), 879–919.

Driscoll, M., Nikula, J., & DePipper, J. N. (2016). *Mathematical thinking and communication: Access for English learners.* Portsmouth, NH: Heinemann.

Dweck, C. S. (2008). *Mindset: The new psychology of success.* New York: Random House Digital Inc.

Fives, H., & Buehl, M. M. (2012). Spring cleaning for the "messy" construct of teachers' beliefs: What are they? Which have been examined? What can they tell us? In *APA educational psychology handbook, Vol 2: Individual differences and cultural and contextual factors.* https://doi.org/10.1037/13274-019.

Gallagher, M. A., & Ruiz, M. I. (under review). Changing teachers' knowledge, beliefs, and practices for teaching mathematics to English language learners.

Garet, M. S., Porter, A. C., Desimone, L., Birman, B. F., & Yoon, K. S. (2001). What makes professional development effective? Results from a national sample of teachers. *American Educational Research Journal, 38*(4), 915–945. https://doi.org/10.3102/00028312038004915.

Gomez Zwiep, S. S., & Benken, B. B. (2013). Exploring teachers' knowledge and perceptions across mathematics and science through content-rich learning experiences in a professional development setting. *International Journal of Science & Mathematics Education, 11*(2), 299–324. https://doi.org/10.1007/s10763-012-9334-3.

Hansen-Thomas, H. (2009). Reform-oriented mathematics in three 6th grade classes: How teachers draw in ELLs to academic discourse. *Journal of Language, Identity, and Education, 8*(2–3), 88–106.

Izsák, A., Tillema, E., & Tunç-Pekkan, Z. (2008). Teaching and learning fraction addition on number lines. *Journal for Research in Mathematics Education, 39*(1), 33–62.

Khisty, L. L., & Chaval, K. B. (2002). Pedagogic discourse and equity in mathematics: When teachers' talk matters. *Mathematics Education Research Journal, 14*(3), 154–168.

Krashen, S. D., & Terrell, T. D. (1983). *The natural approach: Language acquisition in the classroom.* San Francisco, CA: The Alemany Press.

Lamon, S. J. (1993). Ratio and proportion: Connecting content and children's thinking. *Journal for Research in Mathematics Education, 24*(1), 41–61. https://doi.org/10.2307/749385.

Lempp, J. (2017). *Math workshop: Five steps to implementing guided math, learning stations, reflection, and more, grades K-6.* Sausalito, CA: Math Solutions.

Loucks-Horsley, S., Stiles, K. E., Mundry, S., Love, N., & Hewson, P. W. (2009). *Designing professional development for teachers of science and mathematics*. Thousand Oaks, CA: Corwin Press.
Lucas, T. (2011). Language, schooling, and the preparation of teachers for linguistic diversity. In T. Lucas (Ed.), *Teacher preparation for linguistically diverse classrooms: A resource for teacher educators* (pp. 3–17). New York, NY: Routledge.
Lucas, T., & Grinberg, J. (2008). Responding to the linguistic reality of mainstream classrooms: Preparing all teachers to teach English language learners. In M. Cochran-Smith, S. Feiman-Nemser, & J. McIntyre (Eds.), *Handbook of research on teacher education: Enduring issues in changing contexts* (3rd ed., pp. 606–636). Mahwah, NJ: Lawrence Erlbaum.
Miller, J., & Warren, E. (2014). Exploring ESL students' understanding of mathematics in the early years: Factors that make a difference. *Mathematics Education Research Journal, 26*(4), 791–810.
National Council of Teachers of Mathematics. (2014). *Principles to actions: Ensuring mathematical success for all*. Reston, VA: Author.
Neuman, S. B., & Cunningham, L. (2009). The impact of professional development and coaching on early language and literacy instructional practices. *American Educational Research Journal, 46*(2), 532–566. https://doi.org/10.3102/0002831208328088.
Parker, F., Bartell, T. G., & Novak, J. D. (2017). Developing culturally responsive mathematics teachers: Secondary teachers' evolving conceptions of knowing students. *Journal of Mathematics Teacher Education, 20*(4), 385–407. https://doi.org/10.1007/s10857-015-9328-5.
Philipp, R. (2007). Mathematics teachers' beliefs and affect. In F. K. Lester (Ed.), *Second handbook of research on mathematics teaching and learning* (pp. 257–315). Charlotte, NC: Information Age Publishing.
Polly, D., McGee, J., Wang, C., Martin, C., Lambert, R., & Pugalee, D. K. (2015). Linking professional development, teacher outcomes, and student achievement: The case of a learner-centered mathematics program for elementary school teachers. *International Journal of Educational Research, 72*, 26–37.
Ramirez, N. G., & Celedón-Pattichis, S. (2012). Second language development and implications for the mathematics classroom. In S. Celedón-Pattichis & N. G. Ramirez (Eds.), *Beyond good teaching: Advancing mathematics education for ELLs* (pp. 19–38). Reston, VA: National Council of Teachers of Mathematics.

Stanford University. (2014). *How to learn math: For students* [Online module]. Retrieved from https://lagunita.stanford.edu/courses/Education/EDUC115-S/Spring2014/about.

Sztajn, P., Borko, H., & Smith, T. M. (2017). Research on mathematics professional development. In J. Cai (Ed.), *The compendium for research in mathematics education* (pp. 793–823). Reston, VA: National Council for Teachers of Mathematics.

Tatto, M. T., Peck, R., Schwille, J., Bankov, K., Senk, S. L., Rodeiguez, M., & Holdgreve-Resendez, R. (2012). *Policy, practice, and readiness to teach primary and secondary mathematics in 17 countries: Findings from the IEA teacher education and development study in Mathematics* (TEDS-M). http://www.iea.nl/fileadmin/user_upload/Publications/Electronic_versions/TEDS-M_International_Report.pdf.

U.S. Department of Education, National Center for Education Statistics. (2016). *English language learners in public schools.* Retrieved from https://nces.ed.gov/programs/coe/indicator_cgf.asp.

Van Steenbrugge, H., Lesage, E., Valcke, M., & Desoete, A. (2014). Preservice elementary school teachers' knowledge of fractions: A mirror of students' knowledge? *Journal of Curriculum Studies, 46*, 138–161. https://doi.org/10.1080/00220272.2013.839003.

Vanderbilt University. (2017). *Teaching English language learners: Effective instructional practices* [Online module]. Retrieved from https://iris.peabody.vanderbilt.edu/module/ell/cresource/#content.

Viesca, K. M., Joseph, N. M., & Commins, N. L. (2019). A framework for improving the teaching of mathematics to bi/multilingual learners. In L. C. de Oliveira, K. M. Obenchain, R. H. Kenney, & A. W. Oliveira (Eds.), *Teaching the content areas to English language learners in secondary schools* (pp. 135–150). https://doi.org/10.1007/978-3-030-02245-7_9.

5

Improving Algebra Teachers' Ability to Factor in Language: Complementary Expertise in Action

Alandeom W. Oliveira, Carla Meskill, and Sepideh Yasrebi

With the advent of the Common Core State Standards (CCSS), there has been increased emphasis on the importance of language in teaching high school mathematics. In practice, this means transforming the mathematics classroom into a linguistically rich curricular space where verbal activities such as reading, speech, and writing are central to students' mathematics learning experiences. Meeting the linguistic demands of the CCSS can be particularly challenging for algebra teachers, a content area whose highly abstract nature is often viewed as being more amenable to traditional teaching approaches that involve rote learning (practice of standard algorithms for written computation) and mental calculation

A. W. Oliveira (✉) · C. Meskill · S. Yasrebi
State University of New York, Albany, NY, USA
e-mail: aoliveira@albany.edu

C. Meskill
e-mail: cmeskill@albany.edu

S. Yasrebi
e-mail: syasrebi@albany.edu

L. C. de Oliveira and M. Civil (eds.), *Teaching Mathematics to English Language Learners*, https://doi.org/10.1007/978-3-030-48355-5_5

(silent problem-solving). Rather than participation in meaningful discussion and classroom dialogue, students typically engage in thoughtless manipulation of numerical symbols. Taking a more dialogic instructional approach to algebra clearly requires foundational understandings regarding language for communication in tandem with pedagogical expertise.

In an effort to address this issue, this chapter examines the professional development experiences of two high school teachers (a mathematics instructor and a second language specialist) during a year-long program in upstate New York. The teachers taught at an urban school with one of the highest percentages of ELLs in the state, particularly Students of Interrupted Formal Education (SIFEs). Our main goal is to highlight ways that algebra teachers can be prepared to effectively design dialogic curriculum spaces for the study of abstract forms such as equations and polynomials. We examine our focal teachers' professional development experiences as they collaboratively developed, implemented, and reflectively assessed a series of three algebra lessons. We provide a detailed descriptive account of how classroom implementation was interspersed with professional development sessions in which the teachers received training and support in language-focused instructional strategies that specifically support ELLs learning mathematics. In doing so, we seek to show how access to professional development can help algebra teachers design linguistically rich classrooms supportive of ELLs. We first discuss how meeting the linguistic demands of the CCSS requires foundational understandings regarding language and algebra.

Language and Algebra

"Language is a system of signs," argued linguist Ferdinand de Saussure (2011). In other words, language is a system of representation whose structural basis is constituted as an arbitrary relation between a "signifier" (a sound-image, or its typographical equivalent), and a "signified" (the concept or meaning). That is to say, there is no intrinsic or "natural" relationship (neither based on resemblance or causality) between, say, the signifier t-r-e-e and its signified, that is, the mental image of the arboreal entity conjured up by t-r-e-e. The linguistic sign (the totality of signifier and signified) is constituted purely based on social convention. Simply put, the word "tree" evokes an arboreal entity solely because it is transcoded and contextualized as different, say, from the word "free." How does this have bearing on the issue at hand, that is, teaching and learning in algebra classes? Let's start with the following observation: Algebra, first and foremost, is another foreign language and, as such, has to be understood, transcoded, and treated as one. If this is true, then we cannot fail to notice the extent to which ELLs might suffer under the double burden of having to negotiate their way through at least two foreign languages: English and Algebra to succeed at school. Furthermore, the representational relationship between the signifiers and signifieds in algebra is as abstract and unmotivated, if not more, as in any human language. In fact, algebra seems to owe both its universal status and its difficulty precisely to its absolute conceptual abstractness. As abstractions, algebraic forms make metaphoric reference to mathematical ideas (abstract entities, abstract operations, etc.). As emphasized by Lakoff and Núñez (2000), from a cognitive perspective, "algebra is the study of essences" (p. 110), and algebraic structure (notation and symbolism) must be understood in terms of metaphoric representation of mathematical ideas (conceptual systems).

Consistent with the above definitions, many have adopted a purist perspective regarding the nature of algebraic knowledge, viewing it as the instantiation of pure mathematical knowledge strictly through technical mathematical talk (Restivo, 1993). From this purist perspective, algebra is simply a parade of syntactic variations—a set of structural transformations or concatenations of pure thought. In short, the language used to talk about the nature of mathematical knowledge is the language of mathematics itself. However, as emphasized by many philosophers of mathematics, the "math world" is not simply a logical world of abstract forms, signs, and symbols. It is also a world of social practice wherein human beings make epistemic use of not only symbolism, but also ordinary language (social language) beyond the mathematical register (Pimm, 1984). Symbolic communication is only one aspect of the "math world" which also has a linguistically rich social life (Restivo, 1993).

One setting wherein linguistic manifestation of the social life of mathematics is particularly evident is the mathematics classroom. Even in more traditional classrooms in which the teacher stands in front of the room solving problems on the board while students remain in their seats taking notes, discourse is not simply limited to giving voice to mathematical symbols. Rounds (1987) describes how effective mathematics teachers "fill in the blanks between equal signs" through verbalizing explanations and discussing the mathematics behind the symbolic manipulations being transcribed on the board (i.e., providing verbal elaboration to students). A critical component of this discourse is teacher board work. Because students are usually trying to copy whatever the teacher writes on the board while simultaneously listening to verbal explanations, there is potential for a hearing/writing mismatch—problematic situations wherein students are writing down one aspect of a problem while the teacher has already gone ahead talking through the next step. As a result, teacher elaboration may be missed by students (Rounds, 1987). To avoid this, some teachers prefer to say mathematical ideas first and then write them on the board, whereas others write them down first before saying it (these teachers usually step back so students can see what was written on the board), and yet others write and say mathematical ideas simultaneously.

In this chapter, we examine the practices of an algebra teacher in light of this literature and describe how her practices evolve during her participation in a professional development program as she collaborates with a specialist in second language acquisition. Teachers' observations, collaborations, and reflections were coded, categorized, and are here presented by emergent themes. The lessons they taught and the professional development support and processes they experienced along the way constitute the case for this study of complementary expertise as a form of professional development (Yin, 2009).

Teaching Mathematics to ELLs

In addition to understanding the epistemic interplay of algebra and language, a growing literature on teaching academic subjects like algebra to ELLs can support mathematics educators' professional development.

From across the fields of mathematics education and Teaching English to Speakers of Other Languages (TESOL), this literature highlights ways that teachers can effectively support and scaffold ELL's acquisition of the language/content they need to succeed in school. It should be noted that, despite their prominence, some of the endorsed practices are not consensually agreed upon. The following series of recommendations figure prominently in this literature:

1. **Keep Formulaic Language Use at a Minimum**. Most native speakers of English use formulaic language (e.g., phrasal verbs, fixed semantic units, situation-bound utterances, and idioms) in their daily discourse (Miller & Weinert, 1998; Wray, 2013). However, formulaic language use has a tendency to create comprehension barriers for ELLs. Content area teachers in general are encouraged to avoid using formulaic language use in multilingual classrooms.
2. **Develop Lexical Skills for Algebra Content**. Developing receptive skills in algebra classrooms can be challenging for ELLs. Therefore, it is crucial to pre-teach the target vocabulary items and other content-specific words needed for a lesson. Moreover, content area teachers should set realistic goals on the number of new words presented in every session. Schmitt (2000) suggests teaching an average of 10 new vocabulary items per 60-minute lesson. Gairns and Redman (1986) consider an ideal vocabulary load as eight to twelve productive items per lesson. Introduction to new words by class (adjective, adverb, etc.) is also recommended. This recommendation should be taken with caution as educators should refrain from placing too much emphasis on vocabulary. As critics have posited, supporting ELLs' acquisition of content and language requires more than just vocabulary (Moschkovich, 2007).
3. **Develop an Algebra-Specific Dictionary**. Algebra teachers should encourage students to develop their own individual dictionaries as follows: (1) provide the correct form of the algebra-specific vocabulary item; (2) write its part of speech in parenthesis; (3) provide a short definition for the word; (4) encourage entering ELLs to use a bilingual dictionary to look up the word; (5) contextualize the target vocabulary item within a narrative example; and (6) model the word's

pronunciation. ELLs will thereby have a tailored reference to which to refer use to review algebra-specific vocabulary (Zwiers, 2008; Zwiers & Crawford, 2011).

4. **Use Multimodal Anchored Referents.** Incorporating multimodal referents into classroom discourse greatly aids ELLs' comprehension (Meskill, Mossop, & Bates, 1999; Meskill, Nilson, & Oliveira, 2019). Such digital and analog referents or "props" include manipulatives, visuals, illustrations, and the like through which students experience algebraic concepts through multiple modalities (Fuson, Kalchman, & Bransford, 2005).
5. **Promote Cooperative Language Learning.** There are instances when grouping students with similar algebra and language abilities make sense. Generally, however, students seem to benefit more from working in heterogenous groups (Zwiers, 2008; Zwiers & Crawford, 2011) with some initial training to assist pairs in working together. Content-based conversations and activities help students acquire the necessary skills to engage with the material in classroom (Meskill, Mossop, & Bates, 1999). Such conversation may include elaborating and justifying responses, supporting ideas with specific examples, building on or challenging a partner's ideas, paraphrasing a partner's narrative, and synthesizing content. Students need to be closely monitored during group work so as to ensure that all have opportunities to engage in meaningful algebra-specific conversations.
6. **Set Language Norms.** It is important that all students realize classroom expectations. When norms are set and processes established, all students can benefit from a well-structured, well-managed, and balanced classroom environment (Echevarria & Graves, 2007; Vaughn & Bos, 2012). Building routines and norms establishes what is expected regarding assignments, classroom behavior, class member roles, and so on.
7. **Reducing Teacher Talk Time** can provide the students with more opportunities to verbalize their thought processes and engage in a structured pair/group activity. Furthermore, as teachers work through algebraic concepts either in whole-group or small group formats, it is important to repeat target vocabulary items. This supports students observing how language is used algebraically and helps learners to

remember the targeted words. Frequent use of the board (e.g., writing questions, key concepts, etc.) also provides a visual linguistic representation and helps learners trace their own results and processes (Meskill et al., 2019).

8. **Beware of Homophones and Homonyms.** Homophones (*their/they're/their*, *sum/some*, *hole/whole*, etc.) and homonyms (e.g., *row* as a noun and *row* as a verb, etc.), or literal/figurative use of language as well as other subtle semantic, phonetic, and syntactical variations can cause ambiguity and even misinterpretation among ELLs. It is imperative that these potentially difficult algebra-specific content words be introduced prior to the introduction of new content (Nguyen & Cortes, 2013).
9. **Provide ELLs with Sentence Starters** for elaborating their ideas (e.g., *in other words, to put it in another way*), agreeing and disagreeing (e.g., *although I agree with... to a point, it seems more reasonable...*), making a claim (*my point is..., X is important because...*), moving from general to specific claim (e.g., *for instance...*), etc. Teachers of ELLs often maintain a list of sentence starters on the board and seamlessly encourage their use as part of the instructional conversation (Graff, Birkenstein, & Durst, 2006).

The above recommendations were shared with two high school teachers as a part of the professional development program examined in this chapter.

Participants and Their School

Our two participants were Monica, an English to Speakers of Other Languages (ESOL) specialist and Tara, a mathematics teacher. Both taught at the same urban high school in upstate New York State. Tara taught 10th-grade algebra. Although she had prior experience working with ELLs, Tara saw herself primarily as a content expert; she had limited confidence when it came to dealing with issues related to language acquisition. She often consulted with the faculty in the ESOL department to answer questions and deal with any pedagogical difficulties. She stated:

"I have worked with many of the teachers in the ESOL department... I have gone to them in the past for assistance and guidance, and they care of it for me." Monica was chair of the ESOL department. Monica graduated from a Russian university with a degree in Germanic philology and began learning English as a foreign language at ten. She went on to major in English at the university level. The short version of her language learning philosophy is an environmental one with "instruction, plentiful opportunities to practice, accessible input, and an authentic purpose to produce output. So, a teacher needs to create environment for all those conditions to occur." Such a philosophy suggests an ecological perspective regarding content-language integration wherein learning is conceived as environmentally mediated (Van Lier, 2004). Like many language experts, Monica believes that teaching content to ELLs is mainly a matter of designing classroom environments rich in affordances that support learners to interact in meaningful ways that support acquisition.

The high school class where Monica and Tara taught was located in a city with a large population of foreign-born citizens (one of the largest in the state); 16% of the district's students are classified as ELLs. One of the district's many strategies to support immigrant and refugee families is to provide "sheltered" instruction. A sheltered content class here refers to ELL-only classes where language and content are taught intensively at the same time. It is a temporary and transitional learning space designed to make mainstream curriculum accessible (comprehensible) to ELLs by offering a safe, productive, and low-anxiety environment with ample language supports (Fritzen, 2011). Tara's algebra class was composed of eighteen students (seven female and eleven males) with an average age of 17 (14–20). Fifteen of these students were refugees or SLIFEs (Students of Formal or Interrupted Formal Education) (DeCapua & Marshall, 2011). Along with English, there are eight additional languages spoken within the class: Karen (7), Burmese (5), Arabic (1), Bosnian (1), Nepali (1), Urdu (1), Somalia (1), and Vietnamese (1). Their English language proficiency level was either "emerging" or "entering" as measured by their scores on a standardized state test. The teachers' experience and the diversity of their classes were the main selection factors in establishing our focal teaching pair.

Tara described her challenges working with ELLs as follows:

> The key thing is the age disparity…these kids, they need the foundational skills to even get started to go forward. We come in, we start with like numbers, so we write numbers on index cards, and they do it in their native language, and home language, some of them don't know beyond ten in their home language, not in English. So, that's where those students are starting from… at the end of the year, those students are solving two-step equations… they can hold a pencil now and they can seat quietly.

The Professional Development Program

Tara and Monica participated in a year-long professional development program designed to prepare high school teachers to teach mathematics to ELLs. To this end, teachers were offered a summer institute, online training, and school-year support (Meskill et al., 2014). Another of the program's main components was the pairing of practicing educators with complementary expertise—one mathematics and one language—who collaborated to design lessons that were conducive to the simultaneous acquisition of content and language by ELLs (Meskill & Oliveira, 2019). Together, they also looked at materials and concepts, ESOL-specific materials, study guides, images, and models that could support ELLs' mastery of the mathematical concepts that would appear on state tests. They were also introduced to the findings of recent research on sheltered instruction and content-language integrated learning during a summer institute.

An important component of this program was project staff-guided debriefs on video-recorded lessons. Held on school grounds, these debrief sessions followed the recording of each mathematic lesson. During these sessions, participating teachers and a project staff member joint watched the video and respectfully reflected about it. To ensure a non-evaluative and supportive atmosphere, teachers were provided with a video viewing guide based on Rodgers' (2002) reflective cycle, which instructs teachers as follows: "What do you notice? In this phase, try to keep yourself from judging what you see. All judgmental language is out. Ask yourself: Where in the video do I see that? And, What else might those actions (what you are seeing) mean?"

Tara and Monica were then charged with the task of conducting a needs assessment (a critical evaluation of ELLs' needs and available supports) in their school. Based on this assessment they planned, implemented, recorded, and reflectively analyzed two lessons in Tara's sheltered algebra class (Table 5.1). Although they were acquainted with each other, this was the first time that Tara and Monica had an opportunity to collaborate beyond isolated instances of consultation. Their long-term

Table 5.1 Phases of year-long professional development

Phase	Goals/activities
1. Needs assessment	• Gather ELL student data, and teacher schedules with ELL information • Consider district/building initiatives • Record **first lesson**[a] to assess ELLs' needs • Collect samples of student work to support needs assessment
2. Pre-intervention	• Recorded lesson is viewed, and used for reflection vis-à-vis teaching and student learning • Teachers co-develop **second lesson**[a] to be taught based on needs assessment
3. Intervention I	• Collaboratively developed lesson plan is implemented and video-recorded • Samples of student work to be used as evidence of student learning are collected • Recorded lesson is viewed, and used for reflection vis-à-vis teaching and student learning • Teachers co-develop **third lesson**[a] based on assessment of intervention
4. Intervention II	• Collaboratively developed lesson plan is implemented and video-recorded • Samples of student work to be used as evidence of student learning are collected • Recorded lesson is viewed, and used for reflection vis-à-vis teaching and student learning
5. Intervention assessment	• Evaluate data collected as part the intervention plan • Collaboratively written final report (including reflections)

[a]Each lesson lasted for a single class period (approximately 50 minutes)

collaboration lasted throughout the school year, being supervised and documented as were their lessons, reflections, and briefings. Due to a conflict in schedule, Monica was unable to co-teach alongside Tara as they had originally intended. While their lessons were jointly planned and assessed, they were taught only by Tara in her sheltered algebra class.

Teaching Practices Before and After the PD

During the needs assessment phase, Tara and Monica reflectively analyzed a video-recording of a lesson on polynomials. This initial lesson revealed the typical learning experiences of ELLs in Tara's sheltered algebra class. Reflectively reviewing it allowed Tara and Monica to conduct an initial assessment of these ELLs' learning needs, and to identify ways they could be more effectively supported when faced with the task of learning algebra in a new language not spoken at home. This first lesson's main goal was to teach students to add and subtract polynomials combining like terms. Tara began by handing out a daily assignment —a worksheet with the equation $8 + 3\ (n - 4) = 20$, and the inequality $36 < (16 - 4n)$ which students were to solve themselves before Tara did the same on the document camera. This was followed by a lecture where Tara guided students through a set of twelve practice problems meant to illustrate how to simplify expressions and combine like terms (e.g., "Simplify the following expression $2a^2 + 4a - 9 + a - 5a^2 + 12$"). For most of the lesson, Tara stood by the document camera in the back of the room. She occasionally walked around the room to hand out worksheets, check on students' progress, and answer questions. Mini-lectures (talking while writing on a piece of paper being projected in front of the room) were interspersed with periods of silent work. Although some students were sitting in pairs, they worked individually for the most part, save for an occasional question or comparing of notes. This lesson constituted the starting point for Tara and Monica's subsequent collaborations.

Reflective review of the above lesson with feedback from Monica led to a major shift in Tara's teaching approach. This was clearly evident in that the second lesson (co-designed by Tara and Monica) had neither mini-lectures nor teacher-guided problem-solving. Instead, the paired teachers decided to incorporate cooperative learning, something they had studied at the summer institute. With the focus on graphing, their stated goal was: "students would be able to determine and justify if an ordered pair is a solution to a linear equation." In groups, students first engaged in a card-sorting activity where they matched cards with linear equations to cards with their respective solution. This was followed by a second activity whereby groups discussed solutions to linear equations from sets of ordered pairs using their respective graphs. Tara walked around the room, moving from group to group and encouraging students to peer teach.

However, where the goal had been for students to engage in productive conversation about the problems they were solving, they did very little of this. Despite being in groups, students proceeded to spread and sort the cards silently. As Monica commented:

> We need to think what we can do to get them to use more of the language, and to talk about agreeing or disagreeing and figuring out why one of them got one thing and the other got the other one, rather than just doing the math, one of them did that, she [Fatima] just started scribbling, I'm just gonna show you the math, rather than talk about it.

Such an expected pedagogical outcome is consistent with Shaw's (1981) argument that the emergence of a group is a complex endeavor contingent upon a variety of contextual factors, including proximity (e.g., individuals being in the same place at the same time), common fate (being subject to the same consequences), perception (the individuals seeing themselves as a collectivity or unified entity), motivation (e.g., completing a common task and pursuing common goals), reciprocal awareness (being aware of each other's presence), interactivity (coordination of actions or behaviors), and social structure (a set of social roles).

Their joint reflection on the second lesson led to further adjustments. For their third and final lesson, Tara and Monica co-designed a lesson that integrated group work and teacher-led, whole-class discussion. Its main objective was for students "to demonstrate their knowledge of factoring polynomials." Shifting partially back to her prior class structure, Tara began this third lesson with a daily assignment (a worksheet with the two equations $5(n + 4) - 2n = n + 12$, and $\frac{2n}{3} - 4 = 2$) that students solved themselves mostly in silence. However this time, rather than solving the two equations herself, Tara called on individual students to come to the document camera: "*Fatima is going to take over the first question... you heard from me enough, let's see what SHE says about how to do it, so Fatima, you are the teacher here, you got to tell us how to do it.*" Being used for the first time, this student-sharing activity was a strategy used by Tara to encourage students to take a more dialogic approach to the learning of algebra as topic that could also be mastered through group discussion and oral sense-making rather than quiet thinking and silent scribbling (a modality they might be more familiar with). Informed by what she observed while walking around the room, Tara's selection of students was strategic. She called on students who, despite being able to do algebra, were reluctant to talk about it and share their thoughts.

Student-sharing was then followed by teacher-lead discussion about the mathematical term "factor," and an interactive lecture in which she demonstrated how to factor the trinomial $n^2 + 8n - 20$ on the document camera. The lesson then ended with a group activity in which student pairs collaboratively factored a variety of polynomials on a worksheet (progression binomials, trinomials all positives, mixed signs, gcf trinomials, and two perfect squares).

Impact on Teaching

Three main themes were apparent in our analysis of the impact engendered by the professional development program on Tara and Monica's practices: (re)structuring classroom activity, small-group work, and increased attention to language. Each is described and illustrated below.

(Re)Structuring Classroom Activity

Central to Tara's professional development were the continued changes in the structure of her lessons. From the first to the third lesson, Tara went from a strongly teacher-centered structure to a highly student-centered structure and then back to a more balanced structure that was, at times, teacher-centered and, at others, student-centered. A driving force behind these changes was Monica's feedback, input, and peer support. In their reflective analysis of the first lesson, Monica helped Tara recognize that she was exerting a high degree of control over the learning process and that she could encourage students to talk and learn from each other. However, as is common with new educational endeavors, the pendulum seemed to swing too far in the other direction, hence requiring further adjustment. Monica accounted for these noticeable changes as follows:

> Tara tried a variety of techniques in her class: she started with direct teaching, then experimented with grouping, and finally found the method that worked for her. Her first video proved she was an expert in scaffolding and knew how to make content and language accessible to ELLs. Her second video took her out of the comfort zone and allowed her to engage students in group work that gave them more ownership of their learning. Finally, in the last lesson, Tara combined direct teaching, student-to-whole-class teaching, student-to-student teaching, and individual practice…it was clear that Tara was at her "sweet teaching spot" in that [last] lesson.

Tara herself commented positively on her integrated use of direct instruction and group work in the last lesson, describing it as "*the happy medium that I was hoping to find.*" Throughout the history of US education, school teachers have long sought to strike a balance between epistemological product (knowledge) and process (method), between teacher- and student-centeredness, between summative and formative assessment, and between content and language, to name a few. Some have characterized this state of affairs as a metaphorical "pendulum" perennially swinging back and forth without ever reaching stationary equilibrium (Meskill, 2009). Curriculum theorists like Ellis (2004) point out that "balance is seldom achieved in any curriculum, no matter how it is structured" (p. 61) and "the search for curricular balance is never easy and it remains one of the great challenges to any teacher" (p. 106). Solving this conundrum is imperative if mathematics educators are to succeed in their efforts to integrate content and language and effectively design algebra learning experiences linguistically and epistemically accessible ELLs.

Small-Group Work

Throughout her professional development experiences, Tara repeatedly expressed her affinity for direct instruction and discomfort with collaborative learning. However, through trial and reflection, she came to see that both teaching approaches have advantages and hence a place in sheltered algebra instruction:

> I'm most comfortable with direct instruction and working through examples as a class as it allows me to control the classroom. This mode of instruction, however, holds back many of my students as they have to wait for me to catch up to them... and, many of my students are contented to copy and just wait for me to do the work and right it down...
>
> [During the second lesson] lower-level students required constant reassurance and guidance to continue through the activity, this may be due to the grouping of the students... one of the things that the videos presented to me was that [during group work] the students were assisting one another...this is definitely an asset of group work.

Like Tara, school educators in general have increasingly embraced collaborative pedagogical strategies in which students are brought together to experience content areas in small groups and to collaboratively engage in academic discourse. However, recent research also points to numerous challenges associated with establishing productive collaborative learning environments (Oliveira & Sadler, 2008). Rather than straightforward, group activity is in fact highly dynamic and amenable to unforeseen complications and drawbacks (e.g., reticent students who prefer to work alone). As Tara and Monica found out, one effective way of compensating for such drawbacks is to combine group work with other activity structures including direct teaching. Their approach is consistent with many of the principles of Complex Instruction (Cohen, Lotan, Scarloss, & Arellano, 1999; Featherstone et al., 2011).

Increased Attention to Language

Another emergent theme in our data was the increased attention given to language in each subsequent lesson. While previously Tara had focused her lessons almost exclusively on facilitation of conceptual scaffolding, she became increasingly cognizant of the importance of integrating language focusing strategies. This led to the incorporation of several language strategies into her practice, including repetition of vocabulary, questioning to prompt vocabulary response, and deconstructing words for understanding. For instance, during the third lesson, before asking students to complete a set of problems on factoring polynomials, she facilitated a whole-class discussion aimed at answering the question "What does it meant to factor?" which was eventually defined as "( ) ( ) what to multiply together." In the following transcripts, capital letters indicate emphatic tone (used by the teacher to mark important keywords):

> *Tara*: When I say this word right here, [writes "factor" on the document camera], what does that mean?
>
> *Fatima*: Those numbers that we multiply together
>
> *Tara*: Right, so now I have a question [writes "$n^2+8n-20$"] I'm changing the letter today, who can tell me, what's this thing [pointing to the whole expression]?
>
> *Students*: Bi..., trinomial
>
> *Tara*: Trinomial, now, my direction to you is going to be to FACTOR this trinomial, is it ok the key isn't *x*? Instead of an x today, I'm gonna try an *n*. Does it change what we do?
>
> *Students*: No.
>
> *Tara*: No. So, who tell me, what's the gcf [greatest common factor] of this trinomial?
>
> *Student*: One
>
> *Tara*: One, the leading coefficient here is one...[long exchange follows].

Her careful consideration of the linguistic demands of mathematics was also evident in a handout she created for her students where each practice problem is carefully worded so as to familiarize ELLs with the various ways that addition (*add/sum/combine*) and subtraction (*find difference/subtract from/from subtract*) can be expressed in algebra test items:

1. Add $(4n^2 + 3n - 6)$ and $(2n^2 - 8n + 5)$
2. Find the sum of $(2n^2 + 3n - 1)$ and $(5n - 12)$
3. Find the difference of $(2n^2 + 6n - 1)$ and $(n^2 - 8n + 9)$
4. From $(-2n^2 + n + 2)$ subtract $(n^2 + 5n - 5)$
5. Subtract $(3n^2 - 4n + 1)$ from $(n^2 - 2n + 8)$
6. Combine $(2n^2 + 4n + 6)$ and $(3n - 5)$.

In her final report, Tara indicated that she planned to incorporate additional strategies as suggested by Monica and other TESOL specialists with whom she interacted throughout the program, namely sentence frames, short answer/written responses, and justification in words (students verbalize their justifications). Tara's professional development experiences not only helped her gain language teaching expertise (familiarity and skill with language strategies that were previously absent from her teaching repertoire) but also enabled her to recognize the overall value and effectiveness of approaching algebra instruction linguistically, that is, as a pedagogical endeavor that is both epistemic and linguistic in nature.

Conclusion

Effectively teaching algebra to ELLs requires the understanding of what it means to "speak algebra." This includes recognition that even the most abstract algebraic concept can be taken up as talk and thereby learned. As our focal teacher Tara came to realize, mathematics content is comprised of meaning-making systems that extend far beyond de Saussure's symbols. The humanity and complexity of this is expressed in mathematical poetry like the love poem *The Square Root of Three* where the poet Dave Feinberg eloquently morphs the abstract notion of square root into mathematical love (Feinberg, 2008).

References

Cohen, E. G., Lotan, R. A., Scarloss, B. A., & Arellano, A. R. (1999). Complex instruction: Equity in cooperative learning classrooms. *Theory into Practice, 38,* 80–86.

De Saussure, F. (2011). *Course in general linguistics.* New York: Columbia University Press.

DeCapua, A., & Marshall, H. W. (2011). *Breaking new ground: Teaching students with limited or interrupted formal education in U.S. secondary schools.* Ann Arbor, MI: University of Michigan Press.

Echevarria, J., & Graves, A. W. (2007). *Sheltered content instruction: Teaching English language learners with diverse abilities.* Los Angeles, CA: Pearson Allyn and Bacon.

Ellis, A. K. (2004). *Exemplars of curriculum theory.* Larchmont, NY: Eye on Education.

Featherstone, H., Crespo, S., Jilk, L. M., Oslund, J. A., Parks, A. N., & Wood, M. B. (2011). *Smartertogether! Collaboration and equity in the elementary math classroom.* Reston, VA: National Council of Teachers of Mathematics.

Feinberg, D. (2008). *The square root of three.* Retrieved online from the website https://anythingencee.wordpress.com/2011/12/15/the-square-root-of-three-harold-kumar/.

Fritzen, A. (2011). Teaching as sheltering: A metaphorical analysis of sheltered instruction for English language learners. *Curriculum Inquiry, 41*(2), 185–211.

Fuson, K. C., Kalchman, M., & Bransford, J. D. (2005). Mathematical understanding: An introduction. In M. S. Donovan & J. D. Bransford (Eds.), *How students learn mathematics in the classroom* (pp. 217–256). Washington, DC: National Academies Press.

Gairns, R., & Redman, S. (1986). *Working with words: A guide to teaching and learning vocabulary.* Cambridge: Cambridge University Press.

Graff, G., Birkenstein, C., & Durst, R. (2006). *They say, I say: The moves that matters in academic writing.* New York, NY: W. W. Norton.

Lakoff, G., & Núñez, R. E. (20001). *Where mathematics comes from: How the embodied mind brings mathematics into being.* New York, NY: Basic Books.

Meskill, C. (2009). *Teaching and learning in real time: Media, technologies and language acquisition*. Houston, TX: Athelstan.

Meskill, C., Mossop, J., & Bates, R. (1999). Bilingualism, cognitive flexibility, and electronic texts. *Bilingual Research Journal, 23*(2 and 3), 113–124.

Meskill, C., Nilson, J., & Oliveira, A. (2019). Intersections of language, content and multimodalities: Instructional conversations in Mrs. B's Sheltered English Biology classroom. *American Education Research Journal Open*. Retrieved from http://journals.sagepub.com/home/ero.

Meskill, C., & Oliveira, A. W. (2019). Meeting the challenges of English learners by pairing science and language educators. *Research in Science Education, 49*(4), 1025–1040.

Meskill, C., Oliveira, A., Gregory, K., Oliver, G., Reichen, B., & Witmer, P. (2014). Professional development in support of English learner achievement: Science and ESOL educator collaborations. In L. Minaya-Rowe (Ed.), *Effective educational programs, practices, and policies for English learners*. Charlotte, NC: Information Age Press.

Miller, J. E., & Weinert, R. (1998). *Spontaneous spoken language: Syntax and discourse*. Oxford, UK: Clarendon Press.

Moschkovich, J. N. (2007). Beyond words to mathematical content: Assessing English learners in the mathematics classroom. In A. Schoenfeld (Ed.), *Assessing Mathematical Proficiency* (pp. 345–352). New York, NY: Cambridge University Press.

Nguyen, H. T., & Cortes, M. (2013). Teaching mathematics to ELLs: Practical research-based methods and strategies. *Childhood Education, 89*(6), 392–395.

Oliveira, A. W., & Sadler, T. D. (2008). Interactive patterns and conceptual convergence during student collaborations in science. *Journal of Research in Science Teaching, 45*(5), 634–658.

Pimm, D. (1984). *Speaking mathematically: Communication in mathematics classrooms*. New York, NY: Routledge & Kegan Paul.

Restivo, S. (1993). The social life of mathematics. In S. Restivo, J. P. Van Benbegem, & R. Fischer (Eds.), *Math worlds: Philosophical and social studies in mathematics and mathematics education* (pp. 247–278). Albany, NY: State University of New York Press.

Rodgers, C. R. (2002). Seeing student learning: Teacher change and the role of reflection. *Harvard Educational Review, 72*(2), 230–253.

Rounds, P. L. (1987). Characterizing successful classroom discourse for NNS teaching assistant training. *TESOL Quarterly, 21*(4), 643–671.

Schmitt, N. (2000). *Vocabulary in language teaching*. Cambridge: Cambridge University Press.

Shaw, M. E. (1981). *Group Dynamics*: The Psychology of Small Group Behavior (3rd ed.). New York: McGraw-Hill.

Van Lier, L. (2004). *The ecology and semiotics of language learning: A sociocultural perspective*. Boston: Kluwar.

Vaughn, S., & Bos, C. S. (Eds.). (2012). *Research issues in learning disabilities: Theory, methodology, assessment, and ethics*. New York: Springer Science & Business Media.

Wray, A. (2013). Formulaic language. *Language Teaching, 46*(3), 316–334.

Yin, R. (2009). *Case study research: Design and methods*. Los Angeles: Sage.

Zwiers, J. (2008). *Building academic language: Essential practices for content classrooms*. San Francisco: Jossey-Bass.

Zwiers, J., & Crawford, M. (2011). *Academic conversations: Classroom talk that fosters critical thinking and content understandings*. Portland, ME: Stenhouse Publishers.

6

Exploring Professional Development Utilizing Questioning and Effective Mathematics Instruction for Multilingual Students

Sarah A. Roberts

We need to bring about better practices in mathematics classrooms that engage and support multilingual students in meaningful mathematics. One way to help mathematics teachers consider, construct, and engage with these practices is through professional learning experiences. Overall, we know that only a third of mathematics teachers of multilingual students have had professional development (PD) about how to work effectively with multilingual students (Ballantyne, Sanderman, & Levy, 2008). Even when teachers have had such experiences, they do not feel efficacious teaching multilingual students (Ross, 2014). Teachers, like those who participated in the PD described in this chapter, are craving such experiences, and we have few examples of what such PD for mathematics teachers of multilingual students entails (de Araujo, Roberts, Willey, & Zahner, 2018). There has been some research on a small scale about PD for teachers of multilingual students (de Araujo et al., 2018), such as Celedón-Pattichis' (2010) work, where she found

S. A. Roberts (✉)
University of California, Santa Barbara, Santa Barbara, CA, USA
e-mail: sroberts@education.ucsb.edu

L. C. de Oliveira and M. Civil (eds.), *Teaching Mathematics to English Language Learners*, https://doi.org/10.1007/978-3-030-48355-5_6

that teachers need long-term PD when their curriculum changes. By and large, however, mathematics teachers of multilingual students need more PD opportunities in their content areas.

This chapter is framed around two core ideas: (1) four principles of effective multilingual student instruction for mathematics, which goes beyond vocabulary approaches that are often disjointed and incohesive (de Araujo et al., 2018) and (2) teacher questioning. These core ideas are used to examine a semester-longer professional learning opportunity for three seventh-grade mathematics teachers. Using questioning to develop the students' mathematical discourse and the principles to guide the teachers' broadening coherent support of their multilingual students in their mathematics classroom, the goal of this chapter is to understand: What ideas did teachers develop during the PD that they planned to enact in their instruction around the principles and questioning to support multilingual students? This chapter then shares findings, four illustrations of potential practices to support multilingual learners drawn from the PD that unpack the four principles and the teachers' ancillary ideas about questioning. These findings are firmly situated within the PD—they consider the ideas that the teachers *planned to implement* in their own classrooms, those burgeoning ideas that they grew and contemplated within the professional development. This chapter closes with considerations for lessons learned from this data and experiences, as well as suggestions for what others might do with a similar PD.

Questioning and Multilingual Principles

Questioning and multilingual principles provided the overarching framework that guided both the PD and the analysis of this chapter. They are described next.

Questioning

Most agree that questioning is a facet of good mathematics teaching (Martin, 2015). However, many teachers have avoided using questioning, especially with their multilingual students, particularly when they consider using questions that elicit higher-order thinking, out of concern that multilingual students can handle neither the language nor the mathematics. This contributes to a lack of opportunities for multilingual students to engage in high-level mathematics discourse and cognitively demanding mathematics (Iddings, 2005; Planas & Gorgorió, 2004).

Questioning has a number of roles in the classroom. Teachers can ask questions to assess students' learning (McCarthy, Sithole, McCarthy, Cho, & Gyan, 2016), to enable students to explain their thinking (Franke et al., 2009), and to promote a higher level of mathematical thinking (Hong & Choi, 2018). Questioning ideally involves a two-way discussion, where the teacher not only asks questions but also listens to students (Zhang & Patrick, 2012).

Within the research, a number of teachers of multilingual students have exhibited that questioning is a supportive teaching practice. For example, Shein (2012) found that a fifth-grade teacher's use of gestures in tandem with her questioning aided multilingual students in understanding questioning and content. In another study, Celedón-Pattichis and Turner (2012) identified that students developed new discursive habits, such as more extensive solution strategy explanations, when a kindergarten teacher used regular questioning. Questioning can provide potential opportunities for developing discourse practices.

Four Principles for Effective Mathematics Instruction for Multilingual Students

Many teachers would like "tips," "tricks," or "strategies" for working with multilingual students in their mathematics classroom (de Araujo et al., 2018; Roberts, Bianchini, Lee, Hough, & Carpenter, 2017), as if there were a quick fix for multilingual students and as if there were a deficit within multilingual students. An alternative to this approach is a focus on practices that builds on teachers' everyday pedagogy and supports deeper learning to allow teachers not only to produce an action in class but also to reflect on that practice through an adaptive disposition (de Araujo et al., 2018; Roberts et al., 2017).

This chapter uses four key principles of effective mathematics instructional practices for multilingual students, which are compiled from the research on effective instruction for multilingual students. While these principles are appropriate for any students, they are necessary for multilingual students. This framework was originally applied to work with preservice mathematics and science teachers (see Roberts & Bianchini, 2019: Roberts et al., 2017 for more details). These principles draw together multiple scholars' work across mathematics and science education, and these are principles that occur repeatedly when discussing good and key instruction for multilingual students. Others have similarly suggested: providing multilingual learners with challenging or cognitively demand work/tasks (Lee, Llosa, Grapin, Haas, & Goggins, 2019; Moschkovich, 2012; Ramirez & Celedón-Pattichis, 2012); using multilingual students' cultural and linguistic differences as resources not obstacles (Moschkovich, 2012; Ramirez & Celedón-Pattichis, 2012); balancing proficiency and quality of student thinking (Driscoll, Nikula, & DePiper, 2016; Moschkovich, 2012); and providing opportunities to engage in the complex language of the content area, with an expansive, discourse approach to language that occurs with others (Driscoll et al., 2016; Lee et al., 2019: Moschkovich, 2012).

These four principles provided the foundation for the PD described in this chapter. Along with questioning, these four principles framed the planning of the PD; the conversations that we, as facilitators, had with participating teachers about the teaching and learning of mathematics with multilingual students; and the analysis of the data presented in this study. These four key principles reenforce and overlap with one another. They are:

1. Build on and use multilingual students' funds of knowledge and resources;
2. Provide multilingual students with cognitively demanding work;

3. Provide multilingual students opportunities for rich language and literacy exposure and practice; and
4. Identify academic language demands and supports for multilingual students.

The first principle is *build on and use multilingual students' funds of knowledge and resources* (Moll, Amanti, Neff, & Gonzalez, 1992; Moschkovich, 2002). This principle asks teachers to identify, celebrate, and use the knowledge and skills students, their families, and their communities bring to the classroom. Teachers could engage in such practices as recognizing and utilizing their multilingual students' primary languages as resources for learning; encouraging multilingual students to speak in multiple languages, to use different dialects or registers, and/or to work across varying levels of literacies in their production and display of ideas; and incorporating cultural and community resources into their instruction to make content relevant and meaningful.

The second principle, *provide multilingual students with cognitively demanding work* (Tekkumru-Kisa, Stein, & Schunn, 2015; Tobin & Kahle, 1990; Understanding Language, 2013), requires that multilingual students have the opportunity to engage in the same kinds of cognitively demanding work often reserved only for non-multilingual students (Iddings, 2005; Planas & Gorgorió, 2004). Teachers should provide analytical tasks and focus on the practices, concepts, and core ideas outlined in the *Common Core State Standards—Mathematics* (*CCSS-M*; National Governors Association Center for Best Practices, Council of Chief State School Officers; NGA Center, CCSSO, 2010).

The third principle, *provide multilingual students opportunities for rich language and literacy exposure and practice* (Khisty & Chval, 2002; Khisty, McLeod, & Bertilson, 1990; Moschkovich, 2007), attends to the importance of mathematics teachers offering multilingual students the chance to engage in the language of mathematics. Teachers could create opportunities for students to receive comprehensible input, to produce comprehensible output, and to participate in negotiations of meaning that will advance their English language acquisition and mathematics learning. Teachers could also support their students to use multiple modes of communication—speaking, listening, reading, and writing—in activities and assignments.

The fourth principle is *identify academic language demands and supports for multilingual students* (Aguirre & Bunch, 2012; Rosebery & Warren, 2008). This principle asks teachers to attend to the language demands in the tasks they provide multilingual students and to implement appropriate accommodations so that all students can read disciplinary texts, share their ideas and reasoning in whole class and small group discussions, and write about their mathematical thinking. Teachers could support students in learning the language of mathematics; include modeling the target language (e.g., what a conjecture entails); promote all languages and registers with which multilingual students are comfortable; draw from students' funds of knowledge; and use additional scaffolds, such as gestures, graphic organizers, visuals, and demonstrations (see Roberts & Bianchini, 2019, for additional discussion on these principles and their application in a preservice context).

Methods

Context

This PD occurred in California at a junior high school with an enrollment of almost 450 students. Approximately 45% of these students were classified as English learners.[1] The majority of English learners (>85%) at the school were Spanish-speakers.

Participants

Three seventh-grade junior high school mathematics teachers participated: a White male, bilingual Spanish speaker (Mr. Weston); a White female, monolingual English speaker (Ms. Heller); and an Asian female, bilingual Vietnamese speaker (Ms. Kim). Each had 2.5–5 years of

[1]The author acknowledges the term English learners can marginalize linguistically diverse students and prefers the term multilingual students. However, the author also acknowledges that English learners are the term used predominantly in the literature, and it was used in practice to categorize students in schools.

teaching experience. Seventh-grade mathematics in this district was taught in heterogeneous classrooms, for the most part, with a few enrichment classes (similar to honors classes) and a few support classes (for those students struggling in mathematics). These teachers taught all these mathematics classes across the grade level. These seventh-grade teachers encompassed all the teachers at this grade level, and all volunteered to participate (three out of three). The principal had initially suggested that these teachers might be interested, because they were all novice teachers.

Data Collection

The research team video-recorded three PD meetings (2.5 hours, 6 hours, and 2.5 hours; for a total of 11 hours) focused on teacher questioning and the four principles, also collecting the teachers' written work and reflections. The data additionally included four hour-long interviews with each teacher that focused on their questioning practices, interactions with the tasks, experiences working with multilingual students, experiences in the PD, and experiences teaching the PD lessons. This paper draws on the PD videos and associated teacher work as its primary data, with the interviews as ancillary, triangulating data.

Data Analysis

The first step for the analysis was to create verbatim transcripts of the three PDs (all quotations from these PDs are identified in this chapter by line numbers from these transcripts). I then proceeded through two stages of coding related to the two sources of data. The first stage involved coding the PDs by the four principles of effective mathematics instructional practices for multilingual students. I then coded for the teachers' use of questioning in each of coded principles. I looked across these codes for consistencies and inconsistencies in how the teachers discussed each topic within each of the codes and compared the patterns across and within the three PDs for additional consistencies and inconsistencies across the sets of data and the participants.

Overview of the Professional Development

This PD experience was organized so that teachers and facilitators (the author and a graduate student) met approximately monthly during the spring semester. We held one full-day PD and two after school (approximately two-hour) PD meetings. Two teachers attended all meetings, and Ms. Kim missed the last after school meeting.

The general setup of each PD was the following: (1) complete introductory written reflection on practice (either about work we had been doing or about work we would be doing about questioning and the principles; in PD#2 and PD#3, we revisited lessons the teachers taught or previewed the new topic for the day); (2) read about a key topic related to multilingual students and questioning; (3) solve a rich mathematics task and discuss the mathematics as a group; (4) discuss how to implement questioning with this task with the teachers' class, attending specifically to multilingual learners; (5) plan for how to use this task instructionally, particularly how to implement questioning to support multilingual students; and (6) complete closing written reflection on work completed during the PD. The multilingual learner principles and questioning served as a foundation for the work and were discussed in concert with most of the content.

Multilingual learner principles. Table 6.1 provides an overview of which principles were highlighted in which PD session. We provided an overview of all four of the principles in the first PD and then discussed funds of knowledge and how teachers could draw on their students' prior knowledge in their lessons. We knew that this principle might be difficult for teachers to attend to in their lessons (Roberts & Bianchini, 2019), so we wanted to bring it to the teachers' attention soonest so that it would have the most time for teachers to consider and work with across the sessions. The second PD session, we linked our reading, described below, with language-rich opportunities and academic language demands and supports. These two principles often have a lot of overlap, which allowed us to key on using questioning to work with developing specific language opportunities and supports to address some of the demands in the teachers' classrooms. Finally, in the third PD, we focused on the tasks that teachers were using, as described below, and discussed how the

Table 6.1 Multilingual student principle and questioning foci across PDs

	PD #1 (2-hour PD)	PD #2 (Full-day PD)	PD #2 (2-hour PD)
Multilingual student principle focus	• Overview of principles • Funds of knowledge (making links to students' prior experiences)	• Language-rich opportunities • Academic language demands and supports	• Cognitively demanding tasks
Questioning focus	• Overview of questioning • Teachers' current use of questioning • Brainstorm of how questioning can support multilingual students' discourse • Starting to get more thoughtful and proactive about our own questioning—provide teachers with examples of types of questions they could ask in their mathematics classrooms	• Funneling vs. focusing (types of questions), Questions (article)—connect to language-rich opportunities and academic language demands and supports	• Revisit funneling and focusing questioning and how they affected language-rich opportunities teachers attempted to enact from PD#2 • Types of questions (revisited from PD #1) and how to use with their PD#3 lesson

types of questions might affect the cognitive demand of the teachers' task implementation.

Tasks used. We focused each of the three PD meetings around a single mathematics task. These tasks were: (1) "Orange Fizz Experiment" (Georgia Department of Education, 2016), (2) "Sports Bag" (Mathematics Assessment Resource Service, 2015), and (3) "Who's Watching What?" (Illustrative Mathematics, 2017). These were rich tasks in that they had multiple entry and exit points, included multiple representations, and engaged students in reasoning, all qualities of good mathematical tasks (Silver & Stein, 1996). Additionally, they provided multiple opportunities for students to engage with content, to employ language, and to communicate their thinking in meaningful ways (Roberts & Bianchini, 2019). Each task aligned with the teachers' "curriculum" and associated standards. The teachers' district did not use a textbook; instead, their curriculum was a collection of tasks (a "collective") from sources like Georgia Department of Education (2016) and Mathematics Assessment Resource Service (2007–2015), which were aligned to the *CCSS-M* (NGA Center, CCSSO, 2010).

Background questioning information. In each PD meeting, the project team provided teachers with information about questioning, which the teachers read during the PD, as we anticipated that teachers would not have time to read before they came to our meetings. The readings covered such topics as background information on questioning, different types of questions, and the importance of questioning for their multilingual students. For example, in the second PD, the teachers read the National Council of Teachers of Mathematics (NCTM, 2014) *Principles to Actions* "Pose Purposeful Questions," which, in part, attended to funneling and focusing questions. After reading, we discussed the reading, considered how they had been using such questions with their students, and probed the teachers about such topics as how their questioning had varied with multilingual students compared with non-multilingual students. Table 6.1 provides an overview of the different topics we discussed regarding the multilingual student principles and questioning across each of the PDs.

Planning to implement. A key component of each PD was teachers preparing to enact the tasks with their students. We provided teachers with time to solve the tasks with each other, to discuss their solution strategies, and to discuss anticipated student solutions. Teachers then created lesson plans, considering and developing questions they would utilize while implementing the task. We encouraged teachers to consider how they would support multilingual students, asking, for instance, "How will you attend specifically to multilingual students? Identify three questions you will ask and how you anticipate that students will respond. Consider how you will anticipate multilingual students will respond versus non-multilingual students." We also wanted teachers to evaluate how they would enact aspects of the multilingual student principles in their lessons with their questioning, such as to consider cognitive demand. We asked teachers to reflect on whether focusing questions allowed students, particularly multilingual students, to develop higher-level thinking. Teachers finalized their plans during their regular planning time at their school sites.

Illustrations of Principles and Questioning in PD

This section describes the ideas teachers developed during the PD that they planned to enact in their instruction around the principles and questioning to support multilingual students. Table 6.2 provides an overview of each PD session, the focal principle of that session, illustrations of the principle(s) attended to in that PD session, the teachers' attention to questioning in that session, and examples of questioning from that session. Each PD session will now be discussed to flesh out the experiences briefly highlighted in this table.

PD#1—Principle 1: Funds of Knowledge

This first illustration comes from the first PD session after teachers had discussed types of questions and were beginning to solve the Orange Fizz problem (Georgia Department of Education, 2016). This is a ratio problem, and students have to find which concentration has the best taste, comparing orange juice concentrate, carbonated water, and total amount. Teachers found multiple opportunities to *elicit and build from multilingual students' funds of knowledge* (multilingual student Principle 1), suggesting ways to connect to students' prior knowledge and experiences and draw on students' home life experiences.

This problem was similar to other problems that students had already completed. Mr. Weston noted, "my kiddos have done this part-to-part, because we've done a problem with paint. We've done a problem with saltwater" (PD1, Lines 1676–1677). With this previous problem, students in Mr. Weston's class had had the opportunity to taste associated salt solutions. Additionally, his students had solved a similar part-to-part

Table 6.2 Overview of illustrations of practice involving principles and questioning developed in PD

PD #	Focal principle	Illustrations of principle in practice: Example 1	Illustrations of principle in practice: Example 2	Use of questioning	Example(s) of questioning aligned with tasks
1 (2- hour)	Principle 1: Build on and use multilingual students' funds of knowledge and resources	Draw on students' home life experiences through linking the context of a problem to a drink with which many students would have familiarity.	Draw on students' prior knowledge and experiences in class by connecting a new problem to one that students had done previously, using similar realia for the problem.	• Teachers discuss the types of questions that are in the problems	• "Which formula would have the strongest orange taste and why?" • "Which one is the most orangey"
2 (Full day)	Principle 3: Provide multilingual students opportunities for rich language and literacy exposure and practice	Have students write a progressive exit ticket, where students would write about a topic at the beginning of a unit and add to that throughout the unit. Students would develop their ideas mathematically and linguistically across the unit.	Use journals to have students reflect on tasks, such as comparing shapes.	• Teachers discuss questioning within exit ticket • Teachers discuss the types of questions that are in the problems	• "What mistakes did the student make?" • "How could the student's work be improved?" • "What could you ask him?"

PD #	Focal principle	Illustrations of principle in practice		Use of questioning	Example(s) of questioning aligned with tasks
		Example 1	Example 2		
2 (Full day)	Principle 4: Identify academic language demands and supports for multilingual students	Use a variety of questions, and include prepared scaffolds, like sentence frames, for those richer questions that require explanations, justifications, etc.	Use exit tickets to support students' language development. Students could use each other's anonymous language on their exit tickets to make sense of mathematics.	• Teachers discuss the types of questions they ask and types of scaffolds they provide with those questions. • Teachers build questioning into exit tickets. • Teachers discuss the types of questions that are in the problems	• "Can you compare these two things?" • "Can you make up your own?"
3 (2-hour)	Principle 2: Provide multilingual students with cognitively demanding work	Analyze tasks overall using Stein et al. (2000) rubric on cognitive demand.	Discuss which types of questions within a task have which level of cognitive demand for students (e.g., calculating mean versus what might affect the measure of center and why one measure of center is better than another).	• Teachers discuss the types of questions that are in the problems • Teachers discuss the cognitive demand of the questions that are in the problems	• "Do you think you can use this data?" • "Which measure of center is best do you think? • "Would you use median or mean to describe the center?"

ratio problem, and they would be able to draw on this prior knowledge when they solved the Orange Fizz problem. He wanted to make this connection to their prior knowledge and build on their funds of knowledge. The questions he would ask would be more relatable, because students would already have some familiarity with these types of tasks.

Furthermore, the teachers wanted to draw on the students' home life experiences when they taught the Orange Fizz lesson. The teachers decided that they would link the problem to Jamaica, a drink familiar to many of their students. The teachers would follow Mr. Weston's lead and make the different concentrations of the Orange Fizz problem using Jamaica for the students to try during class time. The teachers discussed what this might entail.

Mr. W: So I'm thinking of actually doing that for them again.
Ms. K: To reveal it at the end that they could taste which one is the most orangey?
Mr. W: Yeah or when we're talking about best, we can talk about does more flavor mean better. (PD1, Lines 1868–1872)

The group felt like drawing on this fund of knowledge would make the problem more relatable to the students, possibly engaging them more in mathematics by grounding the problem in a familiar context. Teachers would draw on students' prior knowledge and experiences in class and out of class to frame the questions they used with part-to-part ratios when they taught Orange Fizz.

PD#2—Principle 3: Language-Rich Opportunities

This second illustration comes from the second PD session, early in the meeting, after teachers had discussed their implementation of the Orange Fizz problem from PD#1. Teachers had shared questions they had asked and how they had supported multilingual students during the Orange Fizz problem implementation. We next considered what they would do differently if they were to teach the problem again. The facilitator asked the teachers the following question: "As you look forward, and you look at developing cognitive demand in your class, how do you think about doing that work?...How do you see your role doing that work?" (PD2, Lines 844–847). Mr. Weston linked this conversation with the third principle for effective multilingual student mathematics instruction, *providing students opportunities for rich language and literacy.* He shared that he wanted students to "explain what they're thinking about it....[and] ask them to write about it" (PD2, Lines 934–939).

Teachers then considered providing rich opportunities for speaking and writing by helping students develop from language of ideas to language of display. This conversation began with Ms. Kim discussing her dream for writing with her students:

> [M]y dream, and I don't have this kind of time for it, is to have journals and to actually have students reflect. It could be all kinds of, it can be an open task. It doesn't have to be a summary all the time. It could be, "Can you compare these two things? Can you make up your own?" (PD2, Lines 976–979)

Ms. Kim wanted to get her kids writing and for that writing to go beyond simply writing summaries of whatever the lesson for that day had been. In particular, Ms. Kim wanted to try a different type of writing with her students. Having students do this different type of writing would help students: (1) develop their writing and (2) help Ms. Kim develop a better relationship with her students. Ms. Kim had been able to pilot such writing during her summer school teaching and found that it was positive for both her and her students.

This conversation led to the group working on rethinking the use of an exit ticket for this purpose. An exit ticket is a written prompt that a teacher could give to students before they left the classroom that could solicit a limited amount of feedback about student learning for a given lesson. However, instead of simply giving a ticket out the door at the end of the lesson, the group formulated a *progressive exit ticket* that they could use with their students throughout a unit. This would be a progressive exit ticket in that students would revisit a topic throughout the unit, in this case three-dimensional figures. A teacher could give this exit ticket out four times across the unit to assess student thinking and have their questioning in writing, letting students add their progressing thinking as it developed on the key ideas for the unit. As the students progressed through the unit, they would be able to move from using *language of ideas*, in this case, all the linguistic resources that a student might have to complete the exit ticket, to using *language of display*, that oral or written language that would be presented to a more academic audience (Bunch, 2014). The teacher would have students come back to this exit ticket repeatedly throughout the unit, adding their revised and developed rough draft knowledge and language (Thanheiser & Jansen, 2016).

This progressive exit ticket could help students to develop and see their mathematical and linguistic ideas across a unit. It also put some of the linguistic demands associated with questioning into writing, which, for some students, might create some space for them to work out their thinking. Students would be able to see how they went from identifying geometric objects by their colloquial names to using the mathematical names, like cylinder, rectangular prism, and triangular prism, while also learning about how to find the volume and surface area of those figures. Across the unit, they would develop ideas about how to talk and write about these concepts and hopefully receive feedback about them.

Figure 6.1 provides an example of this progressive exit ticket as Mr. Weston enacted it at the beginning of the Sports Bag lesson (Mathematics Assessment Resource Service, 2015). In the first enactment of the progressive exit ticket, he showed the students a soup can (as an example of a cylinder), a Toblerone bar (as an example of a triangular prism), and a cereal box (as an example of a rectangular prism) and presented the students with the questions for the first time. He planned to revisit and revise the questions throughout the unit. This would mean in his class, every three days or so, he would return the ticket to the students with his comments that reflected what they had learned that day (e.g., A student wrote, "It's round," and he responded, "What do we call a 3-D figure shaped that way?"). He then planned to provide them with a new copy of the ticket at the end of the class, where they could reflect on the new language and content they had developed (e.g., including dimensions and mathematical language). By the end of the unit, he hoped to be able to remove the sentence frames he planned to provide for the last prompt.

What can you say about these three objects?
What are they named?
What other shapes can you see? Draw them?
You should be able to add two sentences to each shape.

Fig. 6.1 Initial draft of a progressive exit ticket from the Sports Bag Lesson

PD#2—Principle 4: Academic Language Demands

This third illustration, again, comes from the second PD session, shortly after the above conversation related to multilingual student Principle 3. The teachers had read the NCTM (2014) reading about "Posing Purposeful Questions" that included information about funneling and focusing questions, and they had engaged in a discussion of how to prepare students to do a different type of mathematics than has been expected of students previously, with the adoption of standards, such as the *CCSS-M* (NGA Center, CCSSO, 2010). The teachers shared some frustration with how to get students into a habit of explaining their thinking, so the group brainstormed ways that they could do this work with their students and collected these ideas on a whiteboard. The facilitator asked:

> How do you help [students] learn what it means to communicate in a math discourse community?....I want students to respond when I say something. So how do we create that sort of community where students are doing all of those things? (PD2, Lines 2122–2123, 2130–2132).

The multilingual learner fourth principle is *identify academic language demands and supports for multilingual students,* and in this case, the teachers focused on developing adequate scaffolds and supports for oral academic language, written academic language, and mathematical representations; attending to vocabulary, syntax, and discourse; and providing various supports, such as sentence frames, as linked to that above question of supporting students to communicate in a mathematics discourse community. Overall, the teachers considered how to get their students to respond to their questioning and how to engage their students with questioning, when the type of questioning they would be using would be new for both them and their students.

They intended to use a variety of question types. Teachers could use probing questions that supported students to elaborate, expand, or articulate their thinking (NCTM, 2014), as well as known answer questions that had one- and two-word answers. However, if teachers used questions that required more justification, they would need to provide students scaffolds. For example, the facilitator suggested, as a teacher, she might say the following:

> For the last 12 weeks of school, we're really going to work on developing good mathematical justifications. So, for your last 12 weeks with the 7th graders, I want you to develop good mathematical justifications…So, we're going to give you…we're going to start out with some sentence frames. Here's maybe our agreed upon start of sentence frame. What's that going to look like? So, here are some key parts; a good mathematical justification will always have these pieces, okay. (PD2, Lines 2208–2218)

The goal would be to develop good mathematical justifications through teacher questioning. Teachers would create accompanying sentence frames to assist students, particularly linguistically diverse students, as they wrote these mathematical justifications, given the teachers' questions. Teachers would support students with specific scaffolds for the constituent parts of a good justification. Figure 6.2 provides an example of such questioning and scaffolds.

1. Describe what you saw on the other drawings that surprised you.

I did not recognize ____________________ on ____________'s drawing.

I did not include _________________________________ on my drawing that

_______________ included on their drawing.

Something else that surprised me when I drew my bag was

____

____.

2. How did lengths in other groups' drawings compare to yours?

A group that had similar lengths was ___________________.

A group that had different lengths was ___________________.

3. What changes would you like to make to your drawings and why?

One change I would make to my drawing is ______________________,

because____.

4. What advice would you give to other people solving this problem?

One piece of advice I would give to another group is ______________________,

because ____.

Fig. 6.2 Questioning with prepared sentence frames for the Sports Bag problem

The teachers decided they could, building on their work with language rich opportunities, execute this through the use of exit tickets. Ms. Heller explained:

> I couldn't help but go back to what you were saying about exit tickets, too, and thinking how great it could be, a valuable thing, as if there was a way to have students to kind of anonymously submit what their definitions were, and for lack of a better, and if the students could kind of up-vote each other and decide which one they think is the most solid explanation. Then, with it being really low risk, because they don't really know whose explanation it was. But, even if they're not at the point where they could do, have a great explanation, at least they could witness, maybe what their other peers are creating. And, then, they could start to be like, "Oh, I see why I thought this one was more solid than mine was." (PD2, Lines 2229–2237)

In this situation, students could submit an exit ticket anonymously and then have the opportunity to see other students' exit tickets. Again, using an exit ticket allowed students to share their responses to questioning in writing and to have an opportunity to revise their "rough draft" mathematical thinking (Thanheiser & Jansen, 2016). This would give them the chance to develop their own language as they made sense of their peers' mathematical thinking and explanations. Students and their teacher could work together to refine their language and mathematics.

PD#3—Principle 2: Cognitively Demanding Tasks

This fourth illustration comes from the third PD session. The teachers completed warm-up reflection questions about their questioning work since PD#2, sharing the current content on which they were working and how they were applying questioning to that work. The third PD connected questioning with our second multilingual student principle: *provide multilingual students with cognitively demanding work.* We provided teachers with the opportunity to develop an understanding of the cognitive demand of their mathematics tasks. During this PD, rather than providing a preselected task for the teachers, as we had in the previous two sessions, we provided more flexibility for the teachers. Instead, we provided a source, a unit from Illustrative Mathematics

(2017), which aligned with their current unit on statistics and probability. This allowed the teachers to select a task that best aligned with their upcoming content.

We framed our discussions in this PD around selecting a task from a unit with which the teachers had familiarity and that aligned with their current instruction. We then focused on cognitive demand for their multilingual students, reminding teachers of the importance of attending to cognitive demand for multilingual students and revisiting that cognitive demand for multilingual students is often lower (Iddings, 2005; Planas & Gorgorió, 2004). The teachers used a rubric of four different types of tasks and their associated cognitive demands from Stein, Smith, Henningsen, and Silver (2000), which identified low-level cognitive demands (memorization tasks and procedures without connections tasks) and high-level cognitive demands (procedures with connections tasks and doing mathematics tasks) to analyze the tasks within the unit. Teachers read through the details of each level of each type of task.

The teachers then looked at two tasks from the unit. The facilitator posed to them: "[Consider], what kind of tasks am I using in my class? And…where do these tasks fall? Or even where do parts of tasks fall?" (PD3, Lines 403–408). For example, Mr. Weston noted the questions associated with one of the problems:

> Here are a couple of graphs that [show] heights of NBA players, backpack weights, and 30 people at family dinner party. And which measure of center is best do you think? Would you use median or mean to describe the center? So, I like that, it sets them up for comparing… I think that sets us up for doing the analysis of that in this task in terms of some questions that are kind of lower level. Like what's the difference in median distance shown… Do you expect variability? Why or why not? (PD3, Lines 549–561)

The group discussed the fact that these questions were in the procedures with connections tasks, as they required some explanations, requiring students to do more higher-level thinking.

Ms. Heller continued with why they would use the task they selected (Lesson 15 in Illustrative Mathematics, Grade 7, Probability and Sampling, Estimating Measures of Center; Lesson 15.3 "Who's Watching What?"). She shared that she liked that the questions required students to estimate, something that sometimes made her students uncomfortable. Additionally, it required them to analyze data, a higher-order skill.

> What I like [is that] it asks them to do some estimation, which I know they feel really uncomfortable with sometimes. And I think, it's asking them to do that…And, then, it makes them have to think about why they might see this data. Like…there's a 60-year old watching this "Learning to Read." Why might that happen? Huh, maybe the little kid was watching with their grandma. Oh, maybe a 60-year old just moved to this country and is learning a different language. "Learn How to Read" [the problem] is actually making our kids think through those things [and this] is something I don't think they get to do too often, and it does this problem solving skills. (PD3, Lines 583–594)

Overall, the two teachers found that they would be able to engage students with rich mathematical work around measures of central tendency. Students would not just calculate these measures of center, but they would also do work around estimation, understanding what the measures of center meant, what might affect the measure of center, and why one measure of center would be better than another. Teachers felt that they would have questions that required more than simply calculating and that they would be able to ask questions that probed students' thinking with analytic questions, creating an opportunity for a higher cognitive demand lesson.

Conclusion

This PD gave teachers an opportunity to engage in sustained conversations around a framework of principles of effective multilingual student mathematics instruction and questioning. The participating teachers considered their instruction in light of this framing and a larger underlying structure for working with multilingual students. The teachers worked together on three lessons in PD, but they also had larger structures to ground their day-to-day work. These were larger practices and mindsets to develop around the resources that their students brought to class and how they could build on those resources to provide students with rich mathematical experiences.

With that said, this work was not easy, and the participating teachers found this work challenging. The tasks that we used in the PDs were not similar to what they generally used in their day-to-day teaching, which was mostly lecturing to their students. That opens up a question of how to engage practicing teachers in "high leverage practices" (e.g., Ball, Sleep, Boerst, & Bass, 2009) once they have started teaching when such practices are outside of their comfort zone of instruction. We worked with novice teachers on this project, and they shared that it was hard and invigorating to do this work. At one point, one of the teachers decided not to come to the last PD. The participant did not agree with using the tasks we were using, and she struggled with some of the reflections that she was experiencing, among other challenges. The participant did decide to teach the final lesson and completed the final interview to share some of this with us. It reiterated to us the challenges of this work. Today, two of the teachers are no longer at the school; however, the remaining teacher is a leader in the district with implementing mathematics language routines to support linguistically diverse students at his school to participate meaningfully in his class' mathematics discourse community. That is not to say that this PD was the cause of this development, but it may have been a stepping-stone in his development.

This chapter focused on the teachers' planned implementation, as there is a lot of important work that takes place while teachers have those initial moments together in a PLC or PD meeting. Often the teachers we worked with told us that this was the first time that they had gotten to sit down together to work for a sustained period of time to think about how to support linguistically diverse students in mathematics. The PD was the genesis of ideas that often did not otherwise have the opportunity to come to light or to have peers consider and provide feedback. We found that teachers did not always carry out these plans in their classrooms, such as in the case of Ms. Kim's journal, but, other times, in the case of Mr. Weston's progressive exit ticket, he was able to try this out with his students and find some success.

Going forward, there are a few suggestions that we have for others who might do similar work. The first is that multilingual students have to be at the center of the work. This may seem like a "duh" message, but it can be easy to make multilingual students an "assumption." If you are not talking about multilingual students all the time, then they are not front and center—they are not a consideration. Next time through, we would break down each PD (cycle) by a principle and focus on them in detail with a particular aspect of questioning more specifically, continually bringing teachers back to this focus. This might be possible, for instance if you did a two-year PD focus, with each principle being a focus over the course of a semester. Additionally, a key limitation of this study was that it only lasted for a single semester, as it was only a pilot PD study, and so it did not provide teachers with long-term support that might have helped them to be more comfortable with a different day-to-day instructional model. We would suggest a longer-term commitment to any PD, as suggested in the literature (e.g., Garet, Porter, Desimone, Birman, & Yoon, 2001).

This PD provided a space for teachers to integrate questioning within a principle-based framework for effective multilingual learner mathematics instruction. Teachers faced challenges and successes in this process, and they developed instruction that they prepared to enact in their classrooms, creating a space of possibility for innovative instruction for supporting their multilingual learners.

Acknowledgements This researcher was supported by a grant from the University of California, Santa Barbara Academic Senate Council on Research and Instructional Resources Faculty Research Grant. The author would like to acknowledge the tireless contributions of the teachers and their students in this study, without whom this work would not be possible.

References

Aguirre, J. M., & Bunch, G. C. (2012). What's language got to do with it? Identifying language demands in mathematics instruction for English language learners. In S. Celedón-Pattichis & N. Ramirez (Eds.), *Beyond good teaching: Advancing mathematics education for ELLs* (pp. 183–194). Reston, VA: National Council of Teachers of Mathematics.

Ball, D. L., Sleep, L., Boerst, T. A., & Bass, H. (2009). Combining the development of practice and the practice of development in teacher education. *The Elementary School Journal, 109*(5), 458–474. https://doi.org/10.1086/596996.

Ballantyne, K. G., Sanderman, A. R., & Levy, J. (2008). *Educating English language learners: Building teacher capacity. Roundtable report*. Washington, DC: National Clearinghouse for English Language Acquisition & Language Instruction Educational Programs.

Bunch, G. C. (2014). The language of ideas and the language of display: Reconceptualizing "academic language" in linguistically diverse classrooms. *International Multilingual Research Journal, 8*(1), 70–86. https://doi.org/10.1080/19313152.2014.852431.

Celedón-Pattichis, S. (2010). Implementing reform curriculum: Voicing the experiences of an ESL/mathematics teacher. *Middle Grades Research Journal, 5*(4), 185–199.

Celedón-Pattichis, S., & Turner, E. E. (2012). "Explícame tu Respuesta": Supporting the development of mathematical discourse in emergent bilingual kindergarten students. *Bilingual Research Journal, 35*(2), 197–216. https://doi.org/10.1080/15235882.2012.703635.

de Araujo, Z. U., Roberts, S. A., Willey, C. J., & Zahner, W. C. (2018). English learners in K–12 mathematics education: A review of the literature. *Review of Educational Research, 88*(6), 879–919. https://doi.org/10.3102/0034654318798093.

Driscoll, M., Nikula, J., & DePiper, J. N. (2016). *Mathematical thinking and communication: Access for English learners.* Portsmouth, NH: Heinemann.

Franke, M. L., Webb, N. M., Chan, A. G., Ing, M., Freund, D., & Battey, D. (2009). Teacher questioning to elicit students' mathematical thinking in elementary school classrooms. *Journal of Teacher Education, 60*(4), 380–392. https://doi.org/10.1177/0022487109339906.

Garet, M. S., Porter, A. C., Desimone, L., Birman, B. F., & Yoon, K. S. (2001). What makes professional development effective? Results from a national sample of teachers. *American Educational Research Journal, 38*(4), 915–945. https://doi.org/10.3102/00028312038004915.

Georgia Department of Education. (2016, July). Orange fizz experiment. *GSE Grade 7—Unit 7: Ratios and proportional relationships* (pp. 27–38). Retrieved from https://www.georgiastandards.org/Georgia-Standards/Frameworks/7th-Math-Unit-3.pdf

Hong, D. S., & Choi, K. M. (2018). Challenges of maintaining cognitive demand during the limit lessons: Understanding one mathematician's class practices. *International Journal of Mathematical Education in Science and Technology*, 1–27. https://doi.org/10.1080/0020739X.2018.1543811.

Iddings, A. C. D. (2005). Linguistic access and participation: English language learners in an English-dominant community of practice. *Bilingual Research Journal, 29*(1), 165–183. https://doi.org/10.1080/15235882.2005.10162829.

Illustrative Mathematics. (2017). Who's watching what? *Lesson 15: Estimating Population Measures of Center*. Retrieved from https://im.openupresources.org/7/teachers/8/15.html

Khisty, L. L., & Chval, K. B. (2002). Pedagogic discourse and equity in mathematics: When teachers' talk matters. *Mathematics Education Research Journal, 14*(3), 154–168. https://doi.org/10.1007/BF03217360.

Khisty, L. L., McLeod, D., & Bertilson, K. (1990). Speaking mathematically in bilingual classrooms: An exploratory study of teacher discourse. In G. Booker, P. Cobb, & T. Mendicutti (Eds.), *Proceedings of the Fourteenth International Conference for the Psychology of Mathematics Education* (Vol. 3, pp. 105–112). Mexico City: CONACYT.

Lee, O., Llosa, L., Grapin, S., Haas, A., & Goggins, M. (2019). Science and language integration with English learners: A conceptual framework guiding instructional materials development. *Science Education, 103*(2), 317–337. https://doi.org/10.1002/sce.21498.

Martin, D. B. (2015). The collective Black and "Principles to Actions". *Journal of Urban Mathematics Education, 8*(1), 17–23.

Mathematics Assessment Resources Service. (2015). *Designing a 3D product in 2D: A sports bag*. Retrieved from map.mathshell.org/lessons.php?unit=7305&collection=8Mathematics Assessment Resource Service (2007–2015). Index of summative tasks. Retrieved from https://www.map.mathshell.org/tasks.php?gradeid=25

McCarthy, P., Sithole, A., McCarthy, P., Cho, J. P., & Gyan, E. (2016). Teacher questioning strategies in mathematical classroom discourse: A case study of two grade eight teachers in Tennessee, USA. *Journal of Education and Practice, 7*(21), 80–89.

Moll, L. C., Amanti, C., Neff, D., & Gonzalez, N. (1992). Funds of knowledge for teaching: Using a qualitative approach to connect homes and schools. *Theory into Practice, 31*(2), 132–141. https://doi.org/10.1080/00405849209543534.

Moschkovich, J. (2002). A situated and sociocultural perspective on bilingual mathematics learners. *Mathematical Thinking and Learning, 4*(2–3), 189–212. https://doi.org/10.1207/S15327833MTL04023_5.

Moschkovich, J. (2007). Using two languages when learning mathematics. *Educational Studies in Mathematics, 64*(2), 121–144. https://doi.org/10.1007/s10649-005-9005-1.

Moschkovich, J. (2012). *Mathematics, the common core, and language: recommendations for mathematics instruction for ELs aligned with the common core.* Retrieved from https://ell.stanford.edu/sites/default/files/pdf/academic-papers/02-JMoschkovich%20Math%20FINAL_bound%20with%20appendix.pdf.

National Council of Teachers of Mathematics (NCTM). (2014). *Principles to actions: Ensuring mathematical success for all*. Reston, VA: NCTM.

National Governors Association Center for Best Practices, Council of Chief State School Officers. (2010). *Common core state standards mathematics.* Washington, DC: National Governors Association Center for Best Practices, Council of Chief State School Officers.

Planas, N., & Gorgorió, N. (2004). Are different students expected to learn norms differently in the mathematics classroom? *Mathematics Education Research Journal, 16*(1), 19–40. https://doi.org/10.1007/BF03217389.

Ramirez, N. G., & Celedón-Pattichis, S. (2012). Second language development and implications for the mathematics classroom. In S. Celedón-Pattichis & N. G. Ramirez (Eds.), *Beyond good teaching: Advancing mathematics education for ELLs* (pp. 19–38). Reston, VA: National Council of Teachers of Mathematics.

Roberts, S. A., & Bianchini, J. A. (2019). Scaffolding preservice science teacher learning of effective English learner instruction: A principle-based lesson cycle. *Innovations in Science Teacher Education, 4*(3). Retrieved from https://

innovations.theaste.org/scaffolding-preservice-science-teacher-learning-of-effective-english-learner-instruction-a-principle-based-lesson-cycle/.

Roberts, S. A., Bianchini, J. A., Lee, J. S., Hough, S., & Carpenter, S. (2017). Developing an adaptive disposition for supporting English language learners in science: A capstone science methods course. In A. Oliveira & M. Weinburgh (Eds.), *Science teacher preparation in content-based second language acquisition* (pp. 79–96). Columbus, OH: Association of Science Teacher Educators.

Rosebery, A. S., & Warren, B. (Eds.). (2008). *Teaching science to English language learners: Building on students' strengths*. Arlington, VA: NSTA Press.

Ross, K. E. (2014). Professional development for practicing mathematics teachers: A critical connection to English language learner students in mainstream USA classrooms. *Journal of Mathematics Teacher Education, 17*(1), 85–100. https://doi.org/10.1007/s10857-013-9250-7.

Shein, P. P. (2012). Seeing with two eyes: A teacher's use of gestures in questioning and revoicing to engage English language learners in the repair of mathematical errors. *Journal for Research in Mathematics Education, 43*(2), 182–222. https://doi.org/10.5951/jresematheduc.43.2.0182.

Silver, E. A., & Stein, M. K. (1996). The QUASAR project: The "revolution of the possible" in mathematics instructional reform in urban middle schools. *Urban Education, 30*(4), 476–521. https://doi.org/10.1177/0042085996030004006.

Stein, M. K., Smith, M. S., Henningsen, M. A., & Silver, E. A. (2000). *Implementing standards-based mathematics instruction: A casebook for professional development*. New York, NY: Teachers College Press.

Tekkumru-Kisa, M., Stein, M. K., & Schunn, C. (2015). A framework for analyzing cognitive demand and content-practices integration: Task analysis guide in science. *Journal of Research in Science Teaching, 52*(5), 659–685. https://doi.org/10.1002/tea.21208.

Thanheiser, E., & Jansen, A. (2016). Inviting prospective teachers to share rough draft mathematical thinking. *Mathematics Teacher Educator, 4*(2), 145–163.

Tobin, K. G., & Kahle, J. B. (1990). *Windows into science classrooms: Problems associated with higher-level cognitive learning*. Bristol, PA: The Falmer Press, Taylor & Francis Group.

Understanding Language. (2013). *Six key principles for ELL instruction.* Retrieved from Stanford University, Graduate School of Education, Understanding Language website http://ell.stanford.edu/content/six-key-principles-ell-instruction.

Zhang, Y., & Patrick, P. (2012). Introducing questioning techniques to pre-service teachers. *Journal of Teacher Education and Educators, 1*(2), 159–184.

7

Engaging ELLs in Mathematics Learning Through Meaning-Making Activities

Gretchen P. Oliver and Karen M. Gregory

The *Common Core State Standards for Mathematical Practices* call for students to use academic language as they engage with mathematical content (National Governors Association Center for Best Practices & Council of Chief State School Officers, 2010). Rather than students simply acquiring a body of knowledge, these standards require students to engage in disciplinary practices and develop disciplinary core ideas. Indeed, the mathematical practices are linguistically dense and necessitate students' ability to demonstrate conceptual understandings in and through language. They also demand a shift from lower levels of thinking (e.g., recalling, reproducing, etc.) to higher-order thinking skills (e.g., evaluating, analyzing, using evidence, synthesizing, etc.). To complicate matters for students who are new to the United States and who are learning English as a new or second language, English language learners

G. P. Oliver (✉) · K. M. Gregory
Clarkson University, Capital Region Campus, Schenectady, NY, USA
e-mail: goliver@clarkson.edu

K. M. Gregory
e-mail: kgregory@clarkson.edu

L. C. de Oliveira and M. Civil (eds.), *Teaching Mathematics to English Language Learners*, https://doi.org/10.1007/978-3-030-48355-5_7

(ELLs) must also work to meet language proficiency standards as they strive to achieve the same grade-level content standards as their native English-speaking peers.

In spite of the challenges that come part and parcel with the mathematical standards, as well as the linguistic load that they carry with them, there are a number of ways that teachers can facilitate students' engagement with mathematics, paying simultaneous attention to language development, conceptual understandings, and analytical practices (i.e., reasoning, higher-order thinking, depths of knowledge). For example, by adhering to an integrated theory of teaching and learning, teachers can develop a necessary "pedagogical coherence" (Heritage, Walqui, & Linquanti, 2015) to support ELLs as they develop their English language proficiency, mathematical understandings, and disciplinary core practices. Such coherence requires shifts in pedagogy to fully develop language, content understandings, and analytical practices. Furthermore, through meaning-making activities, teachers can assist students as they strengthen their language and reasoning by providing them with authentic opportunities to engage in mathematical practices through a process of "doing and talking" math (Cook, MacDonald, Miller, & Lord, 2017).

This chapter will share practical examples from secondary mathematics classrooms where teachers have engaged their ELLs in mathematics learning, language development, and analytical practices through meaning-making experiences and learning activities. Although their settings varied, each of these teachers shifted their pedagogy from a teacher-centered approach, where ELLs were mostly passive recipients of content, to one that was more student-centered in nature (where students became active participants and agents in their learning and language development). Furthermore, in all three cases, the teachers' shift in pedagogy allowed their ELLs the opportunity to develop conceptual understanding first through interactions, and more specifically through participation in mathematical discourse practices (Moschkovich, 2012), rather than through traditional lecture and note-taking. In doing so, ELLs were given access to authentic mathematical content, opportunities to see themselves as capable thinkers and doers, and the ability to engage in mathematical practices as highlighted by the CCSS as they simultaneously developed English language proficiency.

Framing the Issue: Dialogic Teaching and Learning

Dialogic teaching and learning serves as the theoretical foundation for this chapter. Research supports the sociocultural aspects of learning, the notion of language as action, and the relationship between language and cognition (e.g., Bruner, 1978; Halliday, 1975; Vygotsky, 1962). In his seminal work, Bakhtin (1981) contrasted monologic and dialogic classroom discourse, with the former being teacher-dominated and oftentimes following an IRE (initiation, response, evaluation) pattern (Mehan, 1979; Mercer, 1995; Sinclair & Coulthard, 1975), whereas the latter discourse pattern supports and promotes a learner-centered environment that involves authentic exchanges between the teacher and students. Unlike monologic teaching, dialogic teaching promotes an inclusive learning environment where all participants are invited to be members of a learning community (Lyle, 2008). In a dialogic setting, students are afforded opportunities to use both receptive and productive skills and to "stretch" their language (Krashen, 1982; Swain, 2000). In other words, they move beyond language that is familiar and patterned to language that is less familiar and more complex. Furthermore, teachers are able to model academic language (Schleppegrell, 2012), while students are simultaneously learning language, learning about language, and learning through language (Gibbons, 2015).

Traditionally, mathematics teaching and learning has been more monologic in nature, characterized by a teacher-centered approach and lecture-based instruction. In a bygone era when mathematics instruction prioritized rote memorization and recall, instructional practices such as lecture and drills may have seemed to be more efficient. In the new standards era, however, the kinds of mathematical practices that students are expected to engage in require interaction, discussion, and critical thinking. Clearly, this requires a departure from the monologic side of the instructional spectrum and a pressing need to swing the pendulum to dialogic teaching. In fact, the eight core practices of the Common Core State Standards for Mathematics call for a shift in thinking and pedagogy for teachers and learners. These practices require students to explain their thinking, to read deeply, and to argue with evidence persuasively. This is a challenge for all mathematics students, but especially for ELLs, who are simultaneously learning mathematics content and English as an additional language. Table 7.1 presents the eight core practices for CCSS in mathematics.

Linguistic challenges can be associated with all of these mathematical practices. For example: problems may be conceptually dense and contain unfamiliar sentence structures and language functions (e.g., chronological order, cause-effect); problems may contain tier 2 and tier 3 words, words with multiple meanings, colloquial expressions, and/ or synonyms and discipline-specific words and expressions; the use of varying language functions (e.g., justify, explain, agree/disagree); the use of specific mathematical registers, i.e., language and structures implicit to mathematics contexts; and using logical connectors. In addition, several procedural challenges are associated with the core practices, such as: the use of words, numbers, and symbols to present ideas, interpreting a variety of texts (e.g., graphs, diagrams), following multi-step directions; and, interpreting and creating graphs, charts, and models.

Table 7.1 CCSS mathematical core practices

Standards for mathematical practice[a]
MP1. Make sense of problems and persevere in solving them
MP2. Reason abstractly and quantitatively
MP3. Construct viable arguments and critique the reasoning of others
MP4. Model with mathematics
MP5. Use appropriate tools strategically
MP6. Attend to precision language
MP7. Look for and make use of structure
MP8. Look for and express regularity in repeated reasoning

[a] *Source* http://www.corestandards.org/Math/Practice/

Methodology

The classroom vignettes from three high school mathematics classrooms from New York State featured in this chapter show how teachers shifted their pedagogy from a monologic, teacher-centered approach to a dialogic, student-centered approach. Teachers from these schools participated in a US Department of Education grant-funded professional development and research project, "The Technology-Enhanced Multimodal Observation Protocol (MOP) Project."[1] In each context, the participating teachers began with a needs assessment, where they (1) described their pedagogical approach to teaching mathematics, (2) interviewed each of the ELLs in the focal class (chosen by each individual teacher based on number of ELLs), (3) viewed an initial video recording of their instruction (pre-intervention), and (4) participated in a debrief session with their university partner who served as lead field researcher on the project. Following a process of structured reflection (Rodgers, 2002), we identified areas for focus in their teaching practice, specifically where they would like to support language development, conceptual understandings, and analytical practices among their ELLs, with continuous attention to findings from the needs assessment. Thereafter, each teacher attended a professional development workshop, "Doing and Talking Math and Science"[2] produced by the Wisconsin Center for Educational Research (WCER, 2017), and identified the implementation of meaning-making activities into their practice as a significant component of their intervention. This process continued with two additional video recordings and debrief sessions, and from there, the teachers wrote a final report outlining the evolution of their understandings and approaches to teaching mathematics to ELLs. Figure 7.1 shows the process involved in this research and professional development project. As seen in Fig. 7.1, the needs assessment informs the professional development and the interventions continually.

[1]The Technology-Enhanced Multimodal Observation Protocol (MOP) project took place 2012–2017. It was funded through a grant from the US Department of Education's Office of English Language Acquisition. The contents of this chapter do not necessarily represent the policy of the Department of Education, or imply endorsement by the Federal Government.

[2]Source: http://stem4els.wceruw.org/resources.html.

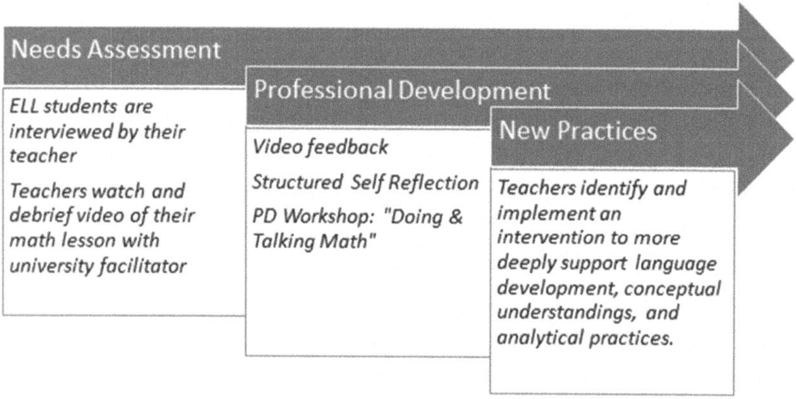

Fig. 7.1 The MOP project process

As shown in each vignette, these teachers, by shifting their pedagogy, provide students with linguistically rich and cognitively challenging opportunities to engage with mathematical content, language, and practices, and in turn, provide them with authentic opportunities to learn with and from one another.

Three Vignettes

Context 1: "Topic in Algebra" Sheltered Course

In this first setting, Mrs. Jones, who teaches mathematics, and her ESL co-teacher, Mrs. Hurst (pseudonyms), have 20 ELL students, speaking eight different home languages, in their introductory Algebra class. Those students' English language proficiency levels range from entering (beginner) to emerging (low intermediate). Several of the students are refugees and have interrupted formal education (i.e., SIFE), resulting in very low math skills. The majority of the students are 9th or 10th graders (14–15 years old), while two are in 11th grade (20 years old), which the teachers explain, "presents challenges in classroom dynamics."

This is a new teaching assignment for both teachers: Mrs. Jones had never taught ELLs, while Mrs. Hurst had never been a co-teacher in a mathematics classroom. Admittedly, at the onset of their participation in the project, both teachers focused on establishing classroom routines and pre-setting students with the requisite mathematical vocabulary that they would need to be able to understand and use in the classroom. Mrs. Hurst explains in the needs assessment narrative,

> We are both new teachers to the course, and the majority of our students are also new to the school, the country, English, and some even to formal education. Our focus is on building classroom routines and introducing basic words and Math skills to our students. Our lessons are mainly teacher-centered. There is not a lot of discourse and many of the students are extremely quiet and timid... we are focused more on content and routine.

After watching an initial video of their co-teaching and participating in the subsequent PD workshop, Mrs. Jones and Mrs. Hurst saw a need to change their approach and find ways to get their students more actively engaged in classroom discourse. Together, they decided to abandon their traditional teacher-centered methods and strategy of "pre-teaching vocabulary," and considered how they would implement more student-directed learning into their pedagogy. Mrs. Jones alludes in a written reflection,

> We feel it is necessary to change the classroom environment to include more discourse and student centered learning. We want students to build on what they already know, with content and language, and explore the concepts we are teaching. This takes time, thought, and focus on our part to find opportunities within the curriculum. We want to push further to discourse initiated by students.

The following vignette shows how they shifted their pedagogy to meet this end.

The 20 students settle into the classroom. Projected on the Smart Board is a "Do Now" activity to review the process of solving a one-step equation: Solve "-4a = 12." Both teachers circulate around the room and assist the students by asking probing questions such as, "What is this operation? What is the opposite of multiplication? What do you need to do to solve for the letter 'a'?"

After all students have had the chance to solve the equation on their paper, Mrs. Hurst asks, "What is the variable?" Students chorally respond, "a." She then asks them, "What operation is associated with 'a?'" Many students respond, "Multiplication." They discuss the other ways they could express multiplication, such as using parentheses or the symbol 'x.' One student solves the equation and tells the class, "a =-3." Ms. Jones then takes them through the process of checking their answer.

The teachers put the 20 students into groups of four and give each group an envelope with 16 index cards. The index cards are color-coded by the following categories: variable, operations, numbers, and equal signs. Mrs. Hurst explains, "You need to work together to put the index cards in an order that shows how to solve the two-step equation." Ms. Jones tells them, "You will see that on the envelope, we have written the answer. As a group, you are going to lay out the cards and solve the two-step equation. You are going to have a discussion and make sure that you all agree. Once you agree, you can tape the index cards to your poster." The teachers also give each group a toolbox (a Ziploc® bag with sentence starters and key vocabulary words to support their conversations).

The students move into their groups, get their materials and begin to work. Each group has a different two-step equation written on their poster paper. As the groups work together, Ms. Jones and Mrs. Hurst check in with the students. Ms. Jones asks one group, "Do you agree? Why? Why not?" Some students respond using the sentence starters, while others simply switch out one index card for another. Mrs. Hurst directs a second group to the example and asks them, "What do you notice? What is similar?" The students talk with one another and place index cards in appropriate spaces as they work through the process of solving the equation. A third group does not agree so Ms. Jones checks in with them to see what they had done. She facilitates a conversation among the four students, asking each one what they would have done, followed by the question, "So, what do we all think?" They come to consensus and complete the two-step equation. Mrs. Hurst sees that a fourth group has solved their equation, and tells them, "Show me how you do the check." Three students take turns in explaining how they check their answer, but one does not participate. Mrs. Hurst reminds them, "Teach him how to do the check and I will come back." She moves to another group and when she returns, that student is able to explain how to solve and check his answer. Once each group has completed the activity and taped the index cards to their poster paper, they present their process to the rest of the class. Ms. Jones and Mrs. Hurst hang the posters in the classroom for all to see.

In a written reflection after the lesson, Ms. Jones commented, "This lesson allowed us to talk with the students and push them to explain their work and thoughts. At first you could tell many students were afraid of this lesson, but just like [Mrs. Hurst] and I stepped out of our comfort zone, we made our students do the same." Likewise, Mrs. Hurst wrote: "The success of the lesson and the presentations were monumental, and students clapped and cheered for each other, even those with the smallest amount of English who could only add something like 'equals 7.' They truly could make meaning by working together through trial and error."

Clearly, the students were "doing and talking math" (Cook et al., 2017) as they engaged in the following math practices: *MP1. Make sense of problems and persevere in solving them*; *MP3. Construct viable and valid arguments from evidence and critique reasoning of others*; *MP5. Use appropriate tools strategically*; *MP6. Attend to precision.* As they worked, they used disciplinary language, including but not limited to *variable, inverse operation, add, subtract, multiply, divide, solution.* At the same time, they also used language registers and structures specific to mathematics, as well as analytical processes that moved them to higher levels of thinking.

Context #2: "Algebra I" Sheltered Course

In this second setting, Mrs. King (pseudonym) teaches Algebra I, a course with a high stakes assessment that is required for graduation, to 18 ELL students. These students range in age from 14 to 20, and the majority of them are in the 10th grade. Similar to the class in the previous context, there are eight home languages spoken here, many of these students are refugees, and a large portion (15 of 18) have been identified as SIFE. Likewise, the proficiency levels of the ELLs in this class are entering and emerging. Mrs. King has created a warm and welcoming classroom environment, and knowing that many of her students have experienced trauma in their home countries, she wants them to feel safe and comfortable. As a result, she prefers to provide them with direct instruction as a way to maintain consistency. At the same time, she sets up her classroom in a way that facilitates peer-to-peer discussion and collaboration.

Mrs. King has taught Math to ELLs in a sheltered setting for several years. While she does not have an ESL co-teacher, she communicates frequently with her colleagues in the ESL department to get information about particular students, as well as to seek advice when she needs assistance and/or guidance. In describing her personal philosophy and pedagogical approach, Mrs. King notes, "I believe that busy students do not create discipline issues. I try and make sure that there is no idle time for them to be unproductive. Guided practice time is essential to make sure that the procedures are followed (correctly). Students feel successful/unthreatened through guided practice." Indeed, this classroom is set up to promote a teacher-centered approach, one where Mrs. King can model mathematical problems using an overhead projector and guide students as they complete problems in a structured, yet highly supported way.

Like her "Topics in Algebra" counterparts, Mrs. King viewed an initial video recording of her instruction and participated in the same "Doing and Talking Math and Science" professional development workshop. She, too, saw a need to make some changes to her practice in order to challenge her students to stretch their language and go deeper in their thinking. In a written reflection, she noted,

> …the video leads me to realize that I need to change up the instructional presentation in the classroom to better meet the needs of all my students. Many of my lower level students are contented [sic] to copy and wait for me to do the work and right [sic] it down. They may know what is needed but are not confident to take that step to get started. They also are not active contributors in a group… This change should allow me to correct some of my shortcomings in differentiating instruction. Through working with my students individually, I am hoping to be able to discover what each student needs to become a better learner. Group work time and allowing students to lead will allow me to work more with individual students on their specific weaknesses and strengths.

As we see in her first reflection, Mrs. King also noticed that her students are at many different levels, both with their language skills and their conceptual understandings of mathematics, and that all students need to be pushed to a level of "productive struggle." As such, she decided to shift to a more collaborative approach, one where her higher-achieving students would be able to serve as leaders, building upon the safe learning environment she created. This change in routine would allow her to give more targeted instruction to support the students who struggle with language and/or content. The following vignette provides an example of how she changed her pedagogy to meet her students' needs.

Eighteen students enter the classroom, and rather than picking up their notebooks and folders to sit and work on a bell ringer as they had in the past, they are placed into groups of three or four, and handed an envelope with a set of index cards: six cards with equations, three cards with integers, and one card with the word "solution" written on it. Mrs. King instructs the students to talk with one another and "sort" the index cards. She asks, "How can you group these cards?" Mrs. King steps back and watches the students for about two minutes. At first, the students are hesitant, yet as leaders in the various groups begin to emerge, conversations slowly begin.

Mrs. King circulates and hands out other materials the students will need to complete the next part of the activity. As she moves throughout the room, she encourages students to talk about why they sort the cards as they do. She asks basic questions such as, "What is the same?" After she has checked in with each group's progress, she brings the class's attention back to her. "Could someone tell me why you put cards together the way that you did?" One student raises his hand and explains how his group sorted the equations from the integers. Mrs. King acknowledges his response, and asks if other groups sorted the cards the same way. Several agree. She then tells the class, "I want you to go back to your groups and think about other ways you can sort the cards. I want you to think like mathematicians. Are there relationships between the different index cards, other than them simply being equations and integers? And how about 'solution?' How can you group the word with the other cards?"

The students turn back to one another and slowly, but silently, they begin to rearrange the index cards. A girl in one group speaks up. She shows the others in her group what she has done. Beginning with the first group "20 = 4n," "3n-6 = 9," and "5," Mrs. King encourages her to explain. She takes her pencil and a piece of paper and does the math, while talking through the process of solving both equations. She moves on to the next set she has grouped, "3-2n = 7," "4n + 9=1" and "n = -2" and does the same. The others nod their heads and a second student takes the pencil so that he can do the math for the next sort. More discussion and conversation follow and all students in the group come to consensus on how to sort the index cards.

Satisfied with their work on this first activity, Mrs. King moves on to the second. She distributes a worksheet with linear equations that need to be matched to solutions, in this case four sets of ordered pairs, as well as a question, "How do you know if the ordered pair is a solution to a linear equation?" A second worksheet has the same linear equations (linear functions), shown on a graph, with two prompts: "Name a solution to the graph;" and "name a point that is ***not*** *a solution to the graph." Rather than employing the standard gradual release of responsibility method (Frey & Fischer, 2008), where she would model her thinking and provide students with guided instruction, Mrs. King positions the students to engage in inquiry through productive group work. She tells the students that they will need to read the directions and determine what the questions are asking them to do, and she releases control of the lesson, allowing students to explore their ideas together. In each of the groups, students begin discussing the questions, and together they work to complete the activity. Additionally, students with strong mathematics and language skills take the lead in their respective groups, which allows Mrs. King to step back and truly see her students' learning in real-time.*

In a written reflection after watching a video recording of this lesson, Mrs. King wrote, "It was enlightening to see how the students interacted in a group setting," and identified the ways the various de facto leaders assisted their peers. She also noted that students were able to use their common L1 to scaffold learning in ways that she could not. Furthermore, she elaborated about the benefits of this type of group work:

> Group work allowed me the opportunity to work more individually with the students. The group setting really enabled differentiated instruction

> to take place. I believe that I was able to better meet the students at their present level and move them forward or at least reinforce the progress they have made, solidifying the knowledge they have already processed. The groups seemed to be most productive for those students who are confident with both their math and language skills.

Mrs. King noted one additional benefit in shifting her pedagogy in this way: She was better able to assess student understanding of both language and math skills. In her final reflection on her videotaped lessons, she wrote,

> The videos would be a lens for me to view my class through the eyes of my students. As an observer, I could view the connections occurring in the classroom, not just the connections to the content but between those in the room, student to student and student to teacher. As the tapings and the debriefings progressed I found myself comforted by the ease of the conversation regarding what was happening in the classroom.

The ELL students Mrs. King's class engaged in several of the core mathematical practices in this activity, for example, *MP1. Make sense of problems and persevere in solving them*; *MP3. Construct viable and valid arguments from evidence and critique reasoning of others*; *MP4. Model with mathematics*; *MP6. Attend to precision.* In this classroom vignette, we see how ELLs not only had opportunities to "do" the mathematics, but also to think and reason, and communicate their thinking and reasoning about the mathematics to one another.

Content #3—Geometry Mainstream Course

This third setting is a co-taught mainstream Geometry class, with both native English-speaking and ELL students, with a high stakes assessment at the conclusion of the course. The math teacher Mr. Barnes is a veteran teacher (22 years) and his ESL co-teacher Ms. Walsh (both pseudonyms) is a first-year teacher. The class has 28 students and includes nine accelerated freshmen, 18 sophomores, and one junior. Additionally, two students have Individualized Education Plans (IEPs) and three are ELLs (all native Chinese-speakers). Furthermore, two of the three ELLs are considered to be strong mathematics students; however, their teachers noted that they struggle with productive language skills.

Through their initial needs assessment, specifically the interviews with their students and viewing the first video recording of their instruction, Mr. Barnes and Ms. Walsh came to a collective understanding that they needed to create opportunities for all students to participate and interact in the mathematics classroom. While the ELLs were more confident and willing to participate in their stand-alone ESL class, they tended to sit quietly and complete their work independently in the mainstream mathematics classroom, and they did not advocate for themselves when they needed help. Therefore, the co-teachers decided that they needed to focus on ways to get the students talking, moving away from teacher-centered instruction, and giving the students some experience with inquiry-based activities. They also sought to find multiple ways to present challenging content, in this case the topic of quadrilaterals through total physical response (TPR) (Asher, 1996) and movement. The rationale for this approach, as they explained, stems from the importance of knowing the properties of these different shapes as they move onto geometric proofs. In their lesson plan, the teachers wrote,

> This is the first day of quadrilaterals [where we will explore] the relationships between the various types and the (introductory) properties of each. Much of the early content students have seen before; still, it is traditionally one of the most difficult topics for students. There are a large number of rote facts students need to know and the content understanding is required at a significantly deeper level than they have ever experienced with these shapes.

The following vignette shows how these teachers introduced quadrilaterals to this class in a way that would allow them to explore the properties of each together through discussion, rather than through a teacher lecture and student note-taking approach.

Ms. Walsh purposefully groups students into fours, and gives each group a very large elastic band. They will use to create various quadrilaterals in order to investigate definitions of quadrilaterals, discover their properties, and deduce the relationship between more specific and less specific classifications. The students move out of the classroom and find space in the quad area, and they are told to move in such a way with the elastic band that they will create the geometric shape that is described to them. First, they will need to listen to Ms. Walsh's descriptions and discuss the shapes they are creating with one another. As she reads the descriptions, Mr. Barnes circulates through the groups, formatively assessing students and asking questions to guide them.

Ms. Walsh tells students, "Make a four-sided figure." In their groups of four, the students move to represent a quadrilateral with the elastic bands. She instructs them to talk with one another about the shape they have made. Some students immediately recognize this as a quadrilateral, while others indicate they have made a specific quadrilateral - a rectangle. They talk about sides and angles, and use the terms "parallel" and "congruent." She continues, "Make a 4-sided figure with both pairs of opposite sides parallel." The students in the groups move. Mr. Barnes asks one group, "Did you need to move?" and then asks another group, "What shape have you made?" He encourages them to share their ideas, which they do. Ms. Walsh commands, "Now, make a 4-sided figure with four congruent sides." One of the groups moves to show a square, while another moves to show a rhombus. Mr. Barnes draws everyone's attention to these two groups and asks, "Are these the same shapes?" Students answer, "No." Mr. Barnes asks, "Which group is correct then?" The students share ideas and one student responds, "They are both correct." He continues to probe their thinking with follow-up questions, engaging them in an instructional conversation (Tharp & Gallimore, 1988).

Next, Ms. Walsh instructs the students to move to show the shapes she calls out, without presenting them with any of the shapes' features. She begins with "trapezoid," and then adds "isosceles trapezoid." The students move. Mr. Barnes circulates and tells the students to talk about the features of each shape. They talk about sides and angles. Next, Ms. Walsh instructs the students to move to form a square and to discuss the features of this shape. The students move, and as they get into place, Mr. Barnes asks "How is this shape different from the last shapes you made?" The students move and talk within their groups.

Finally, the teachers give students a worksheet and tell them they will continue to use the elastic bands to move around and make quadrilaterals, and in addition, they will draw several examples of the same quadrilateral -

with widely varying properties within the constraints of the shape's definition. They emphasize the need for students to talk about the properties and how the various shapes will have some similarities, but others will have distinct features. At this time, both teachers circulate among the groups and monitor their progress.

In a written reflection after watching a video recording of this lesson, the teachers commented on their ELLs' performance. They noted that one of the higher achieving math students stepped into a leadership role and asserted herself more than she would typically do in a more "traditional" lesson. The other two ELLs required additional scaffolding (specifically follow-up questions, prompting, and sentence starters), but they, too, were able to participate more actively. Additionally, Mr. Barnes noticed that the types of questions they asked of the ELLs were different than those questions they asked native English-speaking students. For example, when reviewing his video lesson, Mr. Barnes noticed different patterns of IRE with some students, those whom he may view as weaker and not wanting to back them into an uncomfortable position of being less able to answer. He noted, "I should look for more opportunities to incorporate Talk Moves to move past IRE and facilitate a conversation, not just a response. This will raise the level of language use and student-student communication in my classroom." Finally, Ms. Walsh noticed that the lower-achieving ELLs seemed to be able to draw the quadrilaterals according to their definitions, which was something she struggled with in their previous unit on triangles. When she asked the student about it, the student indicated that having the opportunity to move and see the shapes as they discussed their properties was helpful to her.

This vignette shows multiple examples of students engaging with core mathematical practices, for example, *MP3. Construct viable and valid arguments from evidence and critique reasoning of others*; *MP4. Model with mathematics; MP6. Attend to precision;* and *MP7. Look for and make use of structure.* Herein, we see examples of hands-on and minds-on learning, where students are given opportunities to simultaneously develop their conceptual understandings, analytical processes, and academic language. They have been given agency to co-construct knowledge through discourse and active participation around this topic.

Implications for Practice and Teacher Education

The examples from these vignettes show how educators have shifted their pedagogy in response to what the CCSS, specifically the eight Core Mathematical Practices, call for. Given the linguistic challenges that are embedded in the mathematical practices, all teachers need to see themselves as teachers of language. As noted by Heritage et al. (2015), "If the introduction of college and career ready standards represents a challenge for ELLs, it also presents an opportunity. College and career ready standards imply significant changes in educational practice, in which ELLs learn language and content simultaneously" (p. 3). Therefore, mathematics teachers, both in-service and pre-service, must be prepared to assume the role of both content specialists and language facilitators. For some in-service teachers, this may require an assessment and redesign of current classroom practices.

Pre-service mathematics teachers not only need to learn theory about teaching their content, but also theory related to second language acquisition if they are to be fully prepared to meet the needs of ELLs in today's mathematics classroom. Furthermore, they must develop understandings about dialogic teaching, meaning-making and communicative activities, and the development of academic language along with content learning. While this is no small feat, a pedagogy that responds to the demands of the new standards will be inclusive and equitable, and it will benefit all students.

This project highlights the importance of context-specific, embedded, and on-going professional development for teachers. In each case, we saw that teachers made actionable observations about their own practice by first viewing a model, next watching a recording of themselves teaching, and lastly reflecting and discussing with colleagues and facilitators. In the same way that teachers are encouraged to shift from a monologic model of teaching to a dialogic model, professional development needs to reflect this shift as well. Instead of workshops that are contextually generalized, lecture-based, and formulaic, a shift to models of professional development that require teachers to reflect on their practice, discuss pedagogy with colleagues, and make informed and specific changes, can be impactful both to teachers and the students they teach.

References

Asher, J. (1996). *Learning another language through actions* (5th ed.). Los Gatos, CA: Sky Oaks Productions.

Bakhtin, M. M. (1981). *The dialogic imagination.* Austin: University of Texas Press.

Bruner, J. (1978). The role of dialogue in language acquisition. In A. Sinclair, R. Jarvella, & W. Levelt (Eds.), *The child's conception of language* (pp. 241–256). New York: Springer-Verlag.

Cook, H. G., MacDonald, R., Miller, E., & Lord, S. (2017). STEM discourse: Strengthening reasoning, strengthening language. *WIDA Focus Bulletin.* Wisconsin Center for Education Research. Retrieved from http://stem4els.wceruw.org/resources.html.

Gibbons, P. (2015). *Scaffolding language, scaffolding learning: Teaching English language learners in the mainstream classroom* (2nd ed.). Portsmouth, NH: Heinemann.

Halliday, M. (1975). *Learning how to mean: Explorations in the development of language.* London: Arnold.

Heritage, M., Walqui, A., & Linquanti, R. (2015). *English language learners and the new standards: Developing language, content knowledge, and analytical practices in the classroom.* Cambridge, MA: Harvard Education Press.

Krashen, S. (1982). *Principles and practices in second language acquisition.* Oxford: Pergamon.

Lyle, S. (2008). Dialogic teaching: Discussing theoretical contexts and reviewing evidence from classroom practice. *Language and Education, 22*(3), 222–240. https://doi.org/10.1080/09500780802152499.

Mehan, H. (1979). *Learning lessons: Social organization in the classroom.* Cambridge, MA: Harvard University Press.

Mercer, N. (1995). *The guided construction of knowledge: Talk among teachers and learners.* Clevedon, UK: Multilingual Matters.

Moschkovich, J. (2012). Principles and guidelines for equitable mathematics teaching practices and materials for English language learners. *Journal of Urban Mathematics Education, 6*(1), 45–57.

National Governors Association Center for Best Practices & Council of Chief State School Officers. (2010). *Common Core State Standards for English language arts and literacy in history/social studies, science, and technical subjects.* Retrieved from www.corestandards.org/assets/CCSSI_ELA%20Standards.pdf.

Rodgers, C. R. (2002). Seeing student learning: Teacher change and the role of reflection. *Harvard Educational Review, 72*(2), 230–253.

Schleppegrell, M. J. (2012). Academic language in teaching and learning. *The Elementary School Journal, 112*(3), 409–418.

Sinclair, J., & Coulthard, R. (1975). *Towards an analysis of discourse: The English used by teachers and pupils.* London: Oxford University Press.

Swain, M. (2000). The output hypothesis and beyond: Mediating acquisition through collaborative dialogue. In J. Lantolf (Ed.), *Sociocultural theory and second language learning* (pp. 97–114). Oxford, UK: Oxford University Press.

Tharp, R., & Gallimore, R. (1988). *Rousing minds to life: Teaching, learning and schooling in social context.* New York: Cambridge University Press.

Vygotsky, L. (1962). *Thought and language* (A. Kozulin, Ed. & Trans.). Cambridge, MA: Harvard University Press.

Wisconsin Center for Education Research. (2017). *Doing and Talking Math and Science.* Retrieved from http://stem4els.wceruw.org/.

8

Functional Language Awareness in Secondary Mathematics Classrooms

Holly Hansen-Thomas and Juliet Langman

This chapter highlights the Functional Language Awareness approach, or FLA, which centers on examining and analyzing how teachers of mathematics for Emergent Bilinguals (EBs) attend to and incorporate language into their teaching. As a conceptual tool, the FLA approach assists teachers in working collaboratively with students by scaffolding their participation in learning through focused and explicit attention to language. Providing an overview of the FLA approach by highlighting the important language functions that form the teacher and student discourse in math classes, including (but not limited to) defining and describing, and explaining for problem-solving, this chapter reveals how middle and high school teachers incorporate the Common Core State

H. Hansen-Thomas (✉)
Texas Woman's University, Denton, TX, USA
e-mail: hhansenthomas@twu.edu

J. Langman
Department of Bicultural-Bilingual Studies, UTSA,
San Antonio, TX, USA
e-mail: juliet.langman@utsa.edu

L. C. de Oliveira and M. Civil (eds.), *Teaching Mathematics to English Language Learners*, https://doi.org/10.1007/978-3-030-48355-5_8

Standards for Mathematics that are embedded in language. The chapter also spotlights some key features of language that mathematics teachers can focus on in order to facilitate communication and language use, specifically lexical, syntactic, and discourse-level language features. These features can be supported and promoted through teacher scaffolding. Drawing on discourse from several classrooms, we show how teachers of EBs use language to scaffold their students' learning (of and through language) within this approach.

To foreground this chapter, we pose the question "What is Functional Language Awareness?" and break down the components of the approach as a whole. Functional Language Awareness is an approach that draws together two concepts that have been present in the language teaching field for some time. The first is the notion of *functional language*; the second, the notion of *language awareness.* What that means is that taking a Functional Language Awareness approach considers the purposes or functions of the language we use when we teach, and also encompasses an awareness of the ways in which we are accomplishing these purposes.

Language Functions

Language functions are the purposes for which we use language, or specifically, the intentions behind our (primarily oral) communication. Thinking about language functions can also help us understand how language is used as a teaching and learning tool. In his seminal work *Explorations in the Functions of Language* (1973), Michael Halliday introduced a taxonomy of seven language functions that he argues encompass all of the things we do with language. These include the following functions: Instrumental; Regulatory; Interactional; Personal; Representational; Heuristic; and Interactional. In order to see how a math teacher would apply the concepts, see examples of representative language one would hear in a mathematics class in Table 8.1.

In addition to the broad language functions that Halliday outlined, the mathematics classroom is a context where we can think of language functions related specifically to academics. Academic standards, such as the Common Core State Standards for Mathematics (CCSSM), for

Table 8.1 Halliday's taxonomy for mathematics purposes

Instrumental: (to express needs) "I need a protractor to complete this problem"
Regulatory: (to direct others' behaviors) "Write the following problem on your paper"
Interactional: (to connect to others) "Hi, how are you feeling this class period?"
Personal: (to express feelings, opinions, identity) "I think geometry is the best!"
Representational: (to convey facts and information) "Percent means out of 100"
Heuristic: (to gain knowledge about the world around you) When doing mathematical induction, we follow the method of proof to prove a statement about a number
Imaginative: (to be imaginative, tell stories and jokes) "Why are obtuse angles so depressed? Because they are never *right*!"

Source National Governors Association Center for Best Practices (NGA) and Council of Chief State School Officers (CCSSO) (2010)

example, describe the functions used in the classroom in quite general ways. The CCSSM emphasize mathematical discourse as a part of reasoning skills. The CCSSM center on two types of standards: practice standards and content standards, which can be aligned by grade and content area. For our purposes, we focus on the practice standards for their utility in facilitating language or discourse in the math classroom. There are eight practice standards that move across all grade levels and math concepts.

1. Make sense of problems and persevere in solving them.
2. Reason abstractly and quantitatively.
3. Construct viable arguments and critique the reasoning of others.
4. Model with mathematics.
5. Use appropriate tools strategically.
6. Attend to precision.
7. Look for and make use of structure.
8. Look for and express regularity in repeated reasoning.

Notable about these practice standards is that they focus more on *practices* and *processes*, and less on the specific content of the coursework. What this means is that for students to become proficient, even if it is ultimately only demonstrated on a multiple-choice standardized test, they need to develop disciplinary discourse. Disciplinary discourse is the ability for learners to understand (listen and read) and present (speak and write) their knowledge about content through discipline-specific language—also known as academic language. Academic language differs from social language in that academic language is rooted in school-based concepts that are generally learned in formal settings, whereas social language is dominated by oral language, and topics involved are generally less formal. In the realm of mathematics, the literature has cited other terms, including academic literacy in mathematics, which encompasses mathematical discourse, practices, and proficiency (Moschkovich, 2015). However, a variety of nomenclature is used to describe academic language and literacy within a specific discipline. We maintain that mathematical language subsumes multiple components including literacy, discourse, and others, but for the purposes of the current chapter, we will use the more simple term (mathematical) language, or academic language.

Within the context of the mathematics classroom, academic language functions that are commonly used include the types of processes students are expected to engage in. These include functions such as *defining*, *describing*, and *explaining*, just to name a few. The function of *defining* terms or concepts, i.e., within basic operations, "subtracting is taking away a number from another"; *describing*, such as what we do when we name and discuss repeating patterns with adjectives, i.e., we see a red circle, then a blue square, then a green triangle, and then another red circle. Another common function used is *explaining*, which is often used for problem-solving, where learners have to detail processes step by step using reasoning to uncover a path to a solution, or to outline the reasoning behind a particular set of decisions taken in solving a problem. Certainly, there are many more functions used in math, such as *classifying*, *comparing* and *contrasting*, *inferring meaning*, *hypothesizing*, *predicting*, and *evaluating*. In this chapter, we will illustrate how language functions can be supported through the FLA model using two examples.

Language functions are things we do with language. The way language functions are enacted, however, is through language features. There are multiple language features that teachers can attend to in order to support EBs in mathematics. These include lexical features, which are also known as key vocabulary. Lexical features can be nouns for labeling or supporting vocabulary like adjectives for describing. Another type of feature is syntactic, or ways in which sentences are structured within academic discourse in mathematics (i.e., grammar). These can include the difference between "x divided by y equals..." and "x divided into y (pieces) equals..." where the use of the preposition (by, into) and the form of the verb together can make a different meaning, or a different mathematical term. That is, 12 divided by 4 of course, is 3. But if you have a pizza and you have divided it into 12 pieces for 4 middle schoolers, each student gets *one third* of the pizza, or 3 pieces. It would not be possible, however, to divide 12 pieces into four pieces; it just doesn't make sense! Additionally, there are discourse-level features such as the formulaic expression, "in order to find the X of a Y, you must first use...". A concrete example of this expression might be the following: "in order to find the hypotenuse of a triangle, you must use the Pythagorean theorem." Lexical, syntactic, and discourse-related language features are used in specific ways in the mathematics classroom in other to support students' abilities to define, explain, list steps in a process, among other linguistic functions. Teachers who are able to understand and make strategic use of the language features in support of language functions can facilitate students' language and content knowledge development.

Language Awareness

A growing body of literature suggests that an awareness of language is a critical component in teachers' pedagogical knowledge and practice (Andrews, 2007; Breidbach, Elsner, & Young, 2011; Echevarria, Vogt, & Short, 2013; Zwiers, 2008). Language awareness has been an important and constant feature in language education for decades, but much of the work has not been in content areas such as mathematics. To date, the predominance of teacher language awareness (TLA) research has been conducted in language (including second language or L2) teachers' classrooms (Andrews, 2007; Borg, 2003; Borg & Burns, 2008). The FLA approach supports the call for attention to the needs of ALL content teachers who work with EBs to have training in language awareness (Lindahl, 2013) and especially for teachers of mathematics and related fields (Hansen-Thomas, Langman, & Farias-Sokoloski, 2019).

The language awareness literature suggests that promoting language awareness as a tool can enable mathematics teachers to teach more effectively (Thornbury, 1997), because as teachers become "conscious of and sensitive to language issues at school… [they can] help more of [their] students fulfill their academic potential" (Breidbach et al., 2011, p. 12). Developing language awareness means coming to understand how language operates in the mathematics classroom for purposes of teaching and learning.

Figure 8.1 presents the components of the FLA model. The FLA approach puts *activity*, or what activities are planned to meet a learning objective during a lesson at the core of teaching (see Fig. 8.1). To engage effectively in learning activities means linking the activities to specific language *functions*, which are represented by specific *language features*, that vary according to the language *contexts* in the classroom. Language context

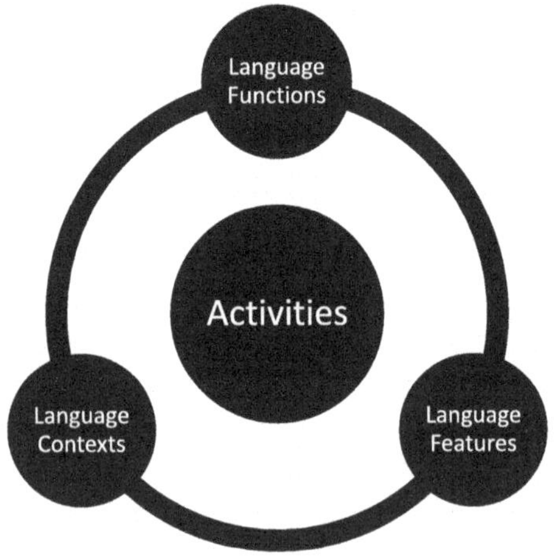

Fig. 8.1 The Functional Language Awareness model

refers to, for example, the different ways we use language to accomplish tasks during lectures, small group brainstorming, and oral or written presentations of work. In a simple way, the language we use to express our ideas can range from informal or everyday language, to academic language associated in this case, with mathematics. The particular constellation of contexts, features, and functions can work to support or hinder students' successful participation in classroom activities. The more a teacher is aware of language, the more she can carefully focus on language as a tool for teaching and learning, the more she or he can support EB learning.

Putting activity at the center of teaching means structuring the classroom to create opportunities for all learners to engage with material in meaningful ways. Teachers teach and students learn through talking—through language, or "languaging." We are using the term "languaging" here in line with the work of Vygotsky and a range of sociocultural theorists of learning. Languaging involves the use of language to do a range of thinking activities: to focus attention, solve problems, to explain, to question, and to create affect (Swain, Lapkin, & Deter, 2013; Swain, Lapkin, Knouzi, Suzuki, & Brooks, 2009). As such, mathematics class is a place in which languaging happens regularly and in multiple ways.

Another way to think about languaging is to think about "scaffolding," a way of providing learners with tools and support for independent learning and problem-solving (Vygotsky & Kozulin, 1986). Scaffolding is a way teachers can structure learning. In broad terms, scaffolding provides a model for teachers to think about what they want their learners to build understanding of during class time. What the FLA model adds to the concept of scaffolding is the conscious attention to the language features that will help to support language functions needed to accomplish an activity in class.

FLA as a Tool for Teachers

The FLA approach provides teachers with concrete tools and strategies for thinking about teaching, and more specifically for thinking about "teacher talk" as a guide and a model for student learning. Teacher talk, or, the *language teachers use in talking with students*, we argue, takes up the largest and most important portion of the teaching day. Most teachers, and in particular those engaged in teaching mathematics, likely have not spent much time thinking about how language serves as a primary tool for teaching and learning. We believe that being provided with a model to think about language as a teaching tool, the FLA model, will help teachers make a sustained shift in their teacher practice.

Discourse Analysis: A Tool for Looking at Language in the Classroom

Discourse analysis is a technique for seeing the connection between the language features we use and the language functions we are fulfilling in different mathematical contexts. Discourse analysis entails looking closely at stretches of language in context, also called discourse. Sometimes discourse analysis is referred to as an analysis of language beyond the sentence level. Classroom discourse analysis involves the examination and analysis of language ongoing in a classroom and is an important tool for reflection (Rymes, 2016). Doing discourse analysis can allow teachers to examine the following aspects of classroom talk: the activity that is occurring between teachers and learners in the classroom; the language functions the activity is aiming to fulfill; the language features being used to fulfill the language function; and the language context in which the language is occurring. By attending to these functions and specific features of language, EBs will have an advantage in comprehending, using, and producing the complex language of math, as well as with the associated mathematical concepts and calculations in class.

The Role of the Teacher

Thus far, we have provided an overview of the FLA approach, including its theoretical underpinnings and primary components. In order to use the FLA approach, the mathematics teacher has certain responsibilities. For starters, it is important that they identify the functions and highlight them by pointing them out to students. They will also share with students the typical language features associated with these functions so students can carry out these functions independently (or work toward independence). Scaffolds for language functions would include specific language features necessary to talk in class. For example, if using the function of defining, the students would need sufficient adjectives to clearly define a term or concept; if explaining, the students would need sentence stems such as "I believe that..." or have an understanding of ordinal (vs cardinal) numbers to be able to enumerate steps in a process (i.e., first, second, third). Teachers will also encourage use of the home language as through use of translanguaging as a pedagogical tool for understanding (Garcia, Ibarra Johnson, & Seltzer, 2017). Additionally, teachers can pay specific attention to their own language (as well as that of their students) by jotting down notes of what is said in the classroom, or even recording some classroom discourse. Then they can examine their language and effectively, conduct an analysis of their classroom discourse to identify where communication and understandings were exceptionally effective, or could have been clearer. In this way, teachers promote and support teacher language awareness.

Examples

To see FLA in action, we can examine the talk ongoing in class by teachers and by students during lessons. Now we will enter a couple of mathematics classes and show some examples of what learners can do with and through language when they help to focus learners' attention on language functions. The examples we present are drawn from classrooms of secondary school mathematics teachers in Texas who work with diverse classes with a range of EBs. We present examples and extrapolate on what these teachers, with whom we have worked over multiple years, do with language in support of developing mathematics knowledge.

Defining in the Classroom

First, we enter Ms. Plane's class 7th grade mathematics class. Her seventh graders are completing a warm-up activity with the aim of reviewing terms and definitions related to geometric shapes, in this case, circles. The pair consisting of Analí and Mari, two EBs who are at different language proficiency levels (Analí at level 2, Emerging, on the WIDA Proficiency Language Development scale (WIDA, 2019), and Mari at level 5, Bridging) are working together. Analí and Mari speak Spanish as their first language or L1 and use both English and Spanish to communicate when working collaboratively in class. The task written on the board asks students to do the following:

Define the terms related to circles.

- Radius
- Diameter
- Circumference.

You will do this in three ways:

1. Tell your partner the definition in your own words.
2. Write a formal definition with words.
3. Write a definition using a visual.

After reading the instructions on the board, Analí and Mari begin to collaborate, with Mari taking the lead.

Example 1: Analí and Mari defining

1. *M*: Okay, so we need to do the words. I will start with radius.
2. *M*: It's, um, it's the distance of the center of a circle to the end of the circle (writing).
3. *A*: End? Ok.
4. *M*: So what is radius?
5. *A*: Pues [well], it's the half.
6. *M*: Huh? Yes, but what do we write?

7. *A*: From point (here) to the end of the circle. Es la mitad del círculo [it's half of the circle].
8. *M*: Ok (writing). It's half of the círculo, I mean, circle! That is a cognate.
9. *M*: There are lots in geometry!
10. *A*: Sí, (whispers to herself) geometría, radio, diámetro…
11. *M*: Yeah. So, then, diameter is the whole length from one end of the circle to another.
12. *M*: I know the formula is two times radius equals the diameter.
13. *A*: Me toca a mí [my turn]. Circumference es la distancia [is the distance] around (draws a circle around a circle)!
14. *M*: Circumference is the distance around a circle. Yes!

In this example, we see how these two learners are working to fulfill the task the teacher has set out for them. They are using features of both English and Spanish, specific vocabulary, as well as syntactic expressions to accomplish the first two task on the board—to write and to share definitions with their partner. We see a combination of formal and informal language between the pair, as they work together, as well as use of two languages. We also see that they are using visuals (line 13) and referring by pointing to materials (line 7) they have already produced in order to accomplish the goal, which is aligned with the language function of defining.

What the teacher has done in providing this task for her learners is to promote use of multiple modalities, having incorporated a variety of tasks (oral and written language, visuals) within the activity assigned. Ms. Plane has an awareness that her learners will benefit from focusing on specific language functions, in this case, defining, and knows that they will need to use not only social language, but also precise academic vocabulary, an important language feature used in defining.

A key CCSSM practice standard we see represented in this segment of discourse is that of "attending to precision." The pair works to construct both oral and written definitions using strategies of translanguaging to utilize their full linguistic repertoires to make meaning (see, for example, García & Wei, 2014; García et al., 2017). Other strategies include attending to cognates, and explicitly focusing on the precise mathematical terms to develop definitions and eventually produce a clear mathematically appropriate definition of each vocabulary item.

In line 2, Mari provides a definition for radius that the pair writes down. Confirming, or possibly questioning Mari's use of the word "end," Analí asks if end was what she said, or what she meant, but accepts the definition and writes it down. Next, in line 4, Mari elicits a response from Analí. Somewhat hesitantly, Mari defines radius as "half" but Mari requests more information as she wants to know what to write for the written response (line 6). Analí then provides a formal definition, using Mari's word "end" from earlier, and then shares further in Spanish what radius is to her—half of the circle (line 7). In line 8, we see that Mari translanguages, speaking both Spanish and English with her partner, but continues writing in English. As an aside, Mari then notes something she has learned about language, namely, the presence of cognates in the geometry lessons they have been working on recently (line 9), which sparks Analí's awareness of the cognates she has seen (line 10). Because Ms. Plane has focused on cognates in the past, the learners pay attention to them. In lines 11 and 12, Mari continues to work on the concept of diameter and provides a definition as well as a formula for this term. In line 13, we see Analí continuing to use translanguaging as a strategy while Mari supports her friend to construct precise language in English for defining and for describing.

As we have seen in the previous example, there are multiple features of language, as well as language moves, or scaffolds, that not only teachers, but learners themselves, can utilize to make meaning through language in math class. Focusing on language as a tool for learning includes use of a variety of tools such as translanguaging, use of cognates, and the repetition of precise definitions. With the students in this class, their full linguistic repertoire meant Spanish and English in both formal and informal forms of language. Cognates, or terms that have the same root in multiple languages, allow students to draw on what they already know about mathematics in this case in another language, in this case Spanish. Repeating precise definitions allows students, particularly EBs, to practice academic English in relatively safe environments, such as peer talk.

Explaining as a Key Step in Problem-Solving

Next, we enter Ms. Carpenter's high school classroom. Here the students are working together to explain how to calculate slope. This activity comes after a series of activities in which Ms. Carpenter has introduced students to slope, provided a demonstration, and has had them work both collaboratively and separately on solving problems involving slope. Here we present an excerpt from a group of three: Ricardo, an EB who speaks Spanish and English and is at level 3: Developing, on the WIDA language proficiency scale; and Kim and Dan, two English-speaking monolinguals. There are a few statements on the overhead projector. The task set forth for the groups is to collaboratively determine if each statement is true or false, and to build an explanation for their reasoning. Finally, they are to be ready to present their answer orally to the class. The statement they are working on is the following:

A line with a slope of 5 is steeper than a line with a slope of ½.

Example 2: Ricardo, Kim, and Dan explaining slope

1. *T*: So, as we start to work these problems, I want you to tell me which is true and which is false and explain why.
2. Work in your small groups.
3. I will come around your groups as you are discussing this.
4. *Ricardo*: I think it's yes, I think it's true.
5. *Kim*: Yeah, me too.
6. *T*: Why do you think it is true? Can you explain why you think so?
7. *Dan*: Well, because it's bigger, the line has a greater slope.
8. *T*: A greater slope? Ok. Ricardo, can you expand on that?
9. *R*: Well, Miss, it's because the line of 5 has greater slope than the line of ½, so it's more steeper.
10. *T*: It is steeper!
11. That's correct Ricardo! Good job.
12. It is because when we are looking at those increasing lines,
13. *Kim*: (reading from her notebook) lines with greater slope have greater changes in Y relative to the same change in X.
14. *R*: Lines with greater slope have greater change in Y...

In this segment, we see the learners working to provide an explanation for their reasoning. The CCSSM practice standard, "Construct viable arguments and critique the reasoning of others," is the most applicable to this excerpt. Since in this example, the students all agree, we do not see them critiquing the reasoning of each other. What we do see, with the scaffolding support of Ms. Carpenter and their notebooks, is how they are working to express a viable argument for their answer. Within this example, the learners and their teacher are incorporating a number of language features to carry out their explanations. These include causal connectives and formulaic expressions. Causal connectives are used commonly in explanations. Terms and phrases that make connections between ideas include: *because* (or *cuz*), *so, then, as a result, consequently*. Formulaic expressions are phrases and partial sentences that occur together on a regular basis. Within the unit on slope, the students have learned many formulaic expressions such as *rise over run, slope equals change in x over change in y*, and the somewhat more specific one we see in this excerpt (lines 13, 14): *lines with greater slope have greater changes in y relative to the same change in x.*

The teacher's scaffolding of explanations is observed in this group interaction. Such scaffolding moves used during the explanation include use of guided questioning, repetition, and prompting. In lines 4 and 5, Ricardo and Kim agree on an answer but have not explained their reasoning. In line 6, Ms. Carpenter comes to their table and prompts them to provide a more elaborated response. Dan then attempts an explanation using a causal connective *because* to explain that the line of 5 has a greater slope than the other line (line 7). The teacher repeats his words "a greater slope" and then prompts through questioning for additional detail (line 8). This time, she directs her questioning to Ricardo, the EB in the group. Ricardo then uses the language they had used in lines 7 and 8 (greater slope) and incorporates the term *steep* (line 9) which has played prominently during the lesson. Ms. Carpenter then models appropriate use of the comparative form of steep (steeper versus more steeper) (line 10) and praises Ricardo's contribution (line 11). Finally, Ms. Carpenter prompts a final statement on slope, providing a sentence starter, "it is because..." (line 12) as a prompt to which first

Kim and then Ricardo respond by reading the formulaic expression from their notebooks (lines 13–14).

Directly following this discussion, Ms. Carpenter realized there was still some discussion about the steepness of the line in the problem the small groups were working on. She brought the students together in a whole class discussion format, referring to Dan's mention about greater slope (line 7) and Ricardo's use of the term "steeper." She then had the whole class reflect on and say aloud: "the greater the slope, the steeper the line" while writing it on the board and drawing a steep line. In so doing, Ms. Carpenter introduced yet another (and admittedly catchier than "lines with greater slope have greater changes in y relative to the same change in x") formulaic expression to facilitate the language function of explaining while exploring the concept of slope.

In the examples, we have illustrated how teachers can use the FLA approach to support the oral academic language of their students in mathematics. In the first example, we showed two EBs working together to construct definitions following a format that the teacher had made explicit in earlier components of that lesson. In the second example, we see how a teacher can scaffold the development of explanations by prompting students to say more, expand on their beliefs, and practice using the language of mathematics as they grapple with the conceptual demands of the subject. What is important to note is that when teachers have an awareness of not only the functions the students will engage in, but also the features, they can specifically plan for instruction of these concepts and support and promote the use of them in math class. In this way, the learners will be equipped with the strategies for independent learning of academic language and mathematical concepts.

Conclusion

The FLA approach is a tool to support student learning and language development and use through an explicit focus on language functions and features embedded within classroom contexts and activities. Given that the CCSSM practice standards call for attention to mathematical practices—many of these being language practices explicit attention to language as a tool for learning is critical for today's mathematics teachers. By attending to specific language functions and features applicable to appropriate standards-based lessons, mathematics teachers using the FLA approach can develop an awareness of their learners' needs and support their learning of academic language and content in mathematics.

References

Andrews, S. (2007). *Teacher language awareness.* Cambridge: Cambridge University Press.

Borg, S. (2003). Teacher cognition in grammar teaching: A literature review. *Language Awareness, 12*(2), 96–108.

Borg, S., & Burns, A. (2008). Integrating grammar in adult TESOL classrooms. *Applied Linguistics, 29*(3), 456–482.

Breidbach, S., Elsner, D., & Young, A. (Eds.). (2011). *Language awareness in teacher education: Cultural-political and social-educational perspectives.* Frankfurt: Peter Lang.

Echevarria, J., Vogt, M., & Short, D. J. (2013). *Making content comprehensible for English learners: The SIOP model* (4th ed.). Boston, MA: Pearson Education.

García, O., Ibarra Johnson, S., & Seltzer, K. (2017). *The translanguaging classroom: Leveraging student bilingualism for learning*. Philadelphia, PA: Caslon.

Garcia, O., & Wei, L. (2014). *Translanguaging: Language, bilingualism and education.* New York, NY: Palgrave Macmillan.

Halliday, M. A. K. (1973). *Explorations in the functions of language.* London: Edward Arnold.

Hansen-Thomas, H., Langman, J., & Farias, T. (2019). The role of language objectives: Strengthening math and science teachers' language awareness with emergent bilinguals in secondary classrooms. *LACLIL, 11*(2), 193–214. https://doi.org/10.5294/laclil.2018.11.2.2.

Lindahl, K. (2013). *Exploring an invisible medium: Teacher language awareness among preservice K12 educators of English language learners.* The University of Utah, ProQuest, UMI Dissertations Publishing.

Moschkovich, J. N. (2015). Academic literacy in mathematics for English learners. *Journal of Mathematical Behavior, 40,* 43–62.

National Governors Association Center for Best Practices, & Council of Chief State School Officers. (2010). *Common core state standards for mathematics*. Washington, DC: National Governors Association Center for Best Practices, Council of Chief State School Officers.

Rymes, B. (2016). *Classroom discourse analysis: A tool for critical reflection* (2nd ed.). New York, NY: Routledge.

Swain, M., Lapkin, S., & Deters, P. (2013). Exploring the effect of languaging activities on cognitive functioning: The case of an older adult in a long-term care facility. *Activities, Adaptation and Aging, 37*(1), 1–18. https://doi.org/10.1080/01924788.2013.760142.

Swain, M., Lapkin, S., Knouzi, I., Suzuki, W., & Brooks, L. (2009). Languaging: University students learn the grammatical concept of voice in French. *The Modern Language Journal, 93*, 5–29.

Thornbury, S. (1997). *About language: Tasks for teachers of English*. Cambridge: Cambridge University Press.

Vygotsky, L., & Kozulin, A. (1986). *Thought and language*. Cambridge: MIT Press.

WIDA. (2019). *ACCESS for ELLs 2.0 interpretive guide for score board*. Regents of the University of Wisconsin System, on behalf of WIDA.

Zwiers, J. (2008). *Building academic language: Essential practices for content classrooms, grades, 5–12*. San Francisco: Jossey -Bass.

9

Promoting Bilingual Mathematical Practices in a Classroom Through Modeling, Roller Coasters, and Discourse

Angela R. Thompson, Michelle Hale, and Alexander Radosavljevic

Teachers are forced to navigate on a daily basis between multiple entities that often oppose each other—coverage of numerous standards, pacing despite students who need remediation, raising test scores, inadequate texts and materials, and what research shows is actually best for student learning. Given these stressful and sometimes crushing demands, teachers are often reluctant to employ practices where they may feel they are relinquishing control over what is taught and learned and become more of a facilitator of student-centered practices (Thompson & Radosavljevic, 2014).

A. R. Thompson (✉)
Governors State University, University Park, IL, USA
e-mail: athompson7@govst.edu

M. Hale
Valley View School District, Romeoville, IL, USA

A. Radosavljevic
Western Governors University, Salt Lake City, UT, USA
e-mail: alex.radosavljevic@wgu.edu

L. C. de Oliveira and M. Civil (eds.), *Teaching Mathematics to English Language Learners*, https://doi.org/10.1007/978-3-030-48355-5_9

While NCTM and other entities (Boaler, 2016; Lesh & Doerr, 2003; Silver, 1994) call for mathematical modeling and authentic tasks in mathematics classrooms, publishing companies have often reduced the idea of modeling to traditional word problems in textbooks. This may be because publishing companies are motivated to say their book is all-inclusive and contains every learning standard required in a subject area by grade level. Administrators look for materials and activities that are "guaranteed," at least statistically, to raise test scores, while at the same time providing "coverage" for all requirements. Using textbook materials as the primary or only source of instruction often makes real-world mathematical connections between students and curriculum seem dry and irrelevant.

To change a child's mindset about their mathematics ability, they need to work on mathematics in an environment that has authentic tasks, engages all students in rigor regardless of their existing knowledge and needs, and creates a purpose for exploring, testing their own ideas, and retesting them when they do not get the desired outcome (National Academies of Sciences, Engineering, and Medicine, 2018). We recognize that mathematics does not exist in a useful way in isolation; it is and needs to be connected to art, engineering, language, culture, discourse, and many other topics. In this chapter, we describe a case study of one teacher's employment of modeling and authentic assessment in her geometry class. Data indicate that the task engaged students in all eight NCTM practice standards, as well as incorporating language and culture as a part of their learning.

The project we describe below is one teacher's response to her principal's request that all teachers use project-based learning (PBL) as a part of their instruction, all year round. According to Eslami and Garver (2013), "PBL can be applied to a classroom in many ways…Projects are central to the curriculum, and present students with a specific problem or challenge, but no specified way of solving it" (p. 122). Tseng, Chang, Lou, and Chen describe project-based STEM learning as "[an approach] that focuses on organizing self-learning in an empirical project. Through practical activities, interactive discussions, …and team cooperation, students reach the planned target and establish their own know-how" (p. 88). We used these definitions to help us understand and explain how our activity may differ from traditional classroom instruction.

Methods

Context for the Study

This study took place at a private Islamic school (K-12) in a suburb of a large Midwest metropolitan city. There are approximately 500 students in the school. Class sizes range from 6 to 21 students. All students in 6–12th grade and all teachers and staff in the building are female. All students are required to take Arabic and religion courses as a part of their study, and daily prayer time is built into the bell schedule. Students wear a uniform and teachers must wear long loose-fitting dresses or skirts with covered legs and arms. About 90% of the students take one or more AP tests, and 100% graduate and go on to college. Families of the students who attend this school are in the middle-class range. The principal and administration promote project-based learning in all classes throughout the school year, though coverage of Common Core State Standards is not required.

Participants

Participants in this study include Miss G. (the teacher) and 16 geometry students in her class. Miss G. is a third-year teacher, licensed by the state to teach secondary mathematics. She is a Caucasian-American, monolingual English speaker. The participants are sophomores and juniors aged 16–18 years old. All but 13% were born in the United States; however, all students are descendants of parents from Middle Eastern countries. All student participants are bilingual English and Arabic speakers. In a survey asking about their language abilities, all the students considered themselves to be fair, good, or excellent in listening, speaking, reading,

Table 9.1 Participants self-reported ability with Arabic language

	Listening (%)	Speaking (%)	Reading (%)	Writing (%)
Minimal	6	6	0	6
Poor	0	31	13	13
Fair	38	19	44	44
Good	50	31	37	31
Excellent	6	13	6	6

and writing in English. Table 9.1 shows the results of the responses of students' self-assessed abilities in each area of Arabic.

All students claim to have learned Arabic either through school, their home country, or family members. Eighty-seven percent of the students speak Arabic and English at home while the other 13% speak only English at home. Most of the participants say they enjoy mathematics and school. The participants worked in groups of four with their chosen group members.

About Miss G.

While Miss G. is relatively new to teaching, she has had an ambitious career so far. She has presented at national and local mathematics education conferences on a number of topics relevant to mathematics teaching and learning. Her idea to have her students create model roller coasters was developed from two needs she identified while reflecting on her practice. First, she noticed that many of her students needed hands-on activities to promote learning, engagement, and retention. Secondly, her school principal asked all teachers to do project-based learning as a regular part of their curriculum. Though her students and most teachers and staff in the school speak Arabic and English, Miss G. only understands and communicates with English.

This year is her third year teaching, and her third time to work with a group of students on the roller coaster project. Some of the modifications and improvements she made based on reflection of previous iterations include: (a) a comprehensive packet for students to make connections between their roller coaster and prior mathematics learning (including volume, dilations, the academic language of geometry, scale factor, trigonometry, estimation, and the use of a clinometer); (b) a revised rubric to provide a detailed description to students on how they would be scored; (c) the addition of a bilingual presentation where students must persuade the audience to invest in their roller coaster design; and (d) several reflection questions so students can share what the experience was like from their perspective.

Miss G. says she has learned a lot about roller coaster construction and can now predict some of the struggles students may encounter. Even so, observations indicate that she allows the students to struggle instead of feeding them the information they need, as long as they continue to be productive in their learning. She has seen most groups experience frustration when their design doesn't work. She explains that the first time a group's marble moves over their construction correctly, it is very exciting to the whole class. "They struggle, then they have a sense of ownership and agency when they get it to work." She also noticed that students who tend to do well in traditional class activities were challenged by the projects, while students who were more often disengaged in traditional class activities were empowered, inspired, and became more like class leaders. In her own words, "high-scoring academics don't do as well [on the project], because it's hands-on, and they aren't used to that. They take comfort in having a method and an answer, and not having to do trial-and-error."

Data Collection and Description of the Project

The project began with a brief visit to the selected geometry classroom to introduce the project and distribute permission forms. Students were told they would design and construct a roller coaster in a group for a marble to "ride" on. Instructions included a limit on the type of materials (construction paper, cardboard, and tape, which were provided), and a list of requirements (e.g., a 360° loop, a funnel, a drop, etc.). The structure of work days was not formalized; each group was free to approach the project in their own ways, and the teacher was available for questions and problems. For the final presentations, participants were instructed to describe and defend their roller coaster and some of their design decisions using both Arabic and English. Every group member was required to speak two languages, though the content of what each said was left up to the students. For context, students were asked to argue for which awards their roller coaster would receive. Awards included scariest ride, tallest, best engineered, and most creative design. Miss G. made it clear that students would not be scored on language accuracy; only that it was important to think and talk about their project in both languages.[1]

Three weeks later, the students received a ten-page packet containing mathematics questions, reflection questions, mathematical calculations, and grading rubrics to help them understand the scope of the project. The project began with a field trip to Six Flags so that they could get roller coaster ideas and think about area, volume, proportional reasoning, and building a model. They were also provided with pictures, videos, examples, and information about building model roller coasters. Almost all class time for the next two weeks was open time to work on projects. However, there were many interruptions due to AP testing, a daily changing bell schedule for Ramadan, and projects from other classrooms. The projects concluded with a brief oral presentation from each group in the third week.

Data includes field notes during work on projects, individual written responses from student packets, field notes from an interview with Miss G., audio recordings of the presentations, video and images of a marble traveling through each completed roller coaster, and a few artifacts from groups.

During construction of projects, participants worked in groups independently, as if they had come to their place of their employment to work with their team. Team members worked, frequently checking in with each other using gestures, objects, and discourse to explain their thinking. Students got supplies by themselves when needed and infrequently asked Miss G. questions. There were no observable classroom management or behavior issues, and students remained on task,

[1]Miss G. and the authors do not read, speak, or write in Arabic, and the participants were told that we would not be able to understand their Arabic discourse.

even assigning "homework" to themselves to prepare for the next day. Discourse of all four groups tended to flow in the same manner:

1. A group member verbalizes an idea
2. Two or more members negotiate that idea with discourse, gestures, and objects
3. The group tests or rejects the idea
4. They discuss and/or modify based on results
5. Repeat

Two groups made a sketch of their roller coasters in advance, and two began cutting, folding, and taping construction paper without a written-out plan. See Fig. 9.1 for a sample sketch. After two weeks, the class did not feel ready, though there was little time left in the school year to provide more work time. The class as a whole elected to come in after school for a review of the final exam so that they could have more class time to work on the roller coasters. Many students also stopped by before and after school, during lunch, and sometimes during prayer time to complete their projects.

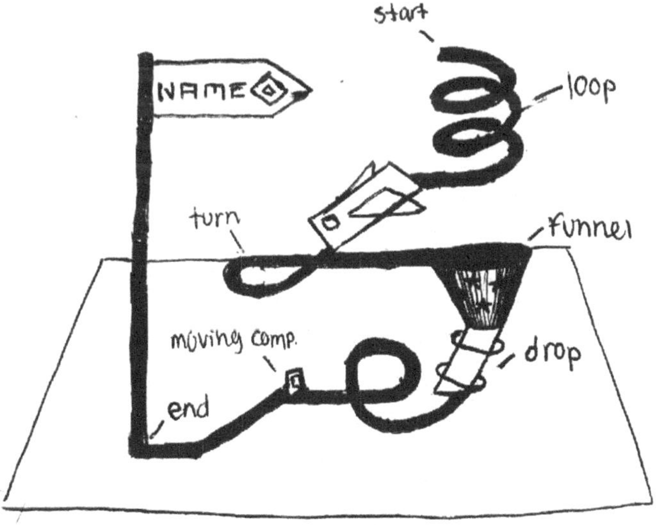

Fig. 9.1 Purple group sketch

Analysis

In the following paragraphs, we describe the uniqueness we observed in each of the four groups, named by the color of construction paper they selected for their project. The description comes from field notes collected during construction of the roller coasters. Field notes were taken during about 75% of the time students worked on the projects. Because some groups were coming in during non-classroom time, and because the researchers were not able to come every day, a few days of observing construction were missed. We are also aware that students were in contact with each other about their projects through texts, emails, and face-to-face interactions outside of the classroom. While we are very excited that students were communicating about their projects outside of class time, we do not have any data from these communications. We focused our analysis of group work time by describing from field notes the uniqueness of each group as they worked together. Our notes were intended to collect evidence of students engaging in the eight NCTM practice standards, which we describe later in this chapter.

Orange Group

Without a closer inspection, some members of this group seemed off task at times. Typically, there were two students working on constructing on the classroom floor, while the other two sat at desks. Upon closer inspection, we observed that the two sitting at desks were not always the same two students, and that the students at desks were completing tasks such as YouTube and Google searches for information, making deep impressions with a pencil on the paper for accurate folding, typing notes in both English and Arabic for the presentation, and cutting tape. This group experienced a tragedy as they came in one day to find out the janitor had disposed of their project completely. They had to start over and rallied together to catch up with other groups. These students had the least amount of discourse between each other; some of their conversation was not about the project, and they frequently worked in near silence. However, they also were clearly working efficiently, often independently, with a common goal. They named their roller coaster *Pumpkin Scare*, to match the orange-and-black color scheme. Of the four groups, this group appears to have actually been the most productive. We feel this way because they were able to catch up despite losing their entire coaster, and their final project was tall, wide, and had many features.

Blue Group

Our field notes labeled this group as blue, but their final product was completely covered with red and white tape to align with their theme. They named their coaster *The Drain*, and decorations were based on the 2017 movie *It*. This group used cardboard rolls to create the tallest and sturdiest frame to attach their roller coaster parts to. They did not have any questions for Miss G. about how to solve problems they may have had; instead, they frequently asked about whether their creative ideas fell within the parameters of the assignment. Much of their discussion was about how to ensure their final product was frightening in numerous ways but safe for the marble. As a part of their final demonstration, one group member played scary music from her phone while the marble traveled its course. Of all the groups, the blue group seemed the most competitive. While they did not imitate ideas or designs from other projects, they worked to earn all nine awards for their project (such as longest ride, tallest ride, most unusual feature, etc.) As a direct result of working on this project, one participant indicated she would like to study engineering. Another student in this group plans to be an interior or architectural designer.

Purple Group

This group's approach was different from the others from start to finish. They had multiple detailed sketches of their plan, though some things changed on the finished product (see Fig. 9.1). Unlike other groups, all four members seem to have spent most of their time problematizing through discourse without much physical evidence of construction. We were concerned that we couldn't see much progress until the last day or two, though we were impressed by their discourse, which we felt sounded like scientists trying to understand a discovery. They unitized the paper they used by drawing gridlines and did not have an external frame as other groups did. Unlike other groups, the marble moved correctly on the track on their very first attempt. The final product was called *The Fruit Loop*, themed after the breakfast cereal. It was remarkably neatly constructed and much smaller when compared with the others. The purple participants described a number of engineering decisions they made during planning, and how those decisions were modified as they worked on producing parts of their model.

Yellow Group

Yellow group experienced the most hardship. In trying to be creative, they began by constructing a tall construction paper frame folded into comparatively large triangular prisms. The frame was about five feet tall. After a week of work, their project had completely collapsed over the weekend and was not repairable. Miss G. guessed that the heavy rain and high humidity caused the weak frame to collapse. They rushed to complete their project, this time with a square-prism frame, but this one also collapsed the day before presentations. The group bonded over this, wondering if their project had been sabotaged, and worked extremely hard with the help of Miss G. in reconstruction. One student stayed after school for hours and placed tracks on the rebuilt frame by herself. As a token to their hardship, they named their coaster *The Malfunctioning Monster*. From our observations, we believe this group learned the most about engineering and construction as they rebuilt their project several times, learning from each iteration. However, the problems they experienced caused them stress and anxiety over completing the project on time. One Yellow member explained in detail how she had learned how *not* to build a roller coaster model.

Results

Based on our data and analysis, we have evidence that all students engaged in all eight NCTM practice standards. In the sections below, we connect the project and our data to each standard. We follow with a discussion on language and assessment, which are critical components to the teaching and implementation of this project.

Mapping Data to Practice Standards

Make Sense of Problems and Persevere in Solving Them

We observed sense-making and perseverance when participants repeatedly tested the marble on their roller coaster (and it didn't work). They proceeded to work together to figure out why parts did not fit together, and how to speed up, slow down, or keep the marble from flying off the tracks. Participants looked for reasons why parts did not work. For example, *Blue Group* decided to have the marble start with a drop and then go directly into a loop. When they tested it, their marble would not make it around the loop, so they decided to make the angle of the drop steeper. Once they did this, the marble dropped at a faster rate and made it all the way around the loop. *Purple Group* talked a lot about how to construct pieces before starting construction. They looked up videos on YouTube or asked each other for advice on how to construct the pieces. In some cases, students would put pieces together, test the marble and it would not land where they needed it to. All groups faced different challenges and contributed to group discourse to solve problems. These groups were able to make sense of their challenges by looking into why it doesn't work and persevere in solving it. A member from *Yellow Group* reflected in her packet "The best part was when the marble worked." Miss G. confirmed that this was the "best part" for groups in the pilot years. What was remarkable was that there was very little evidence of students or groups feeling defeated by the construction not working correctly on a particular trial; we feel this activity meets Boaler's (2016) call for creating a classroom environment where mistakes are valued rather than punished. Students proved very resourceful in working with each other and searching online for solutions to their perceived failures.

Reason Abstractly and Quantitatively

Participants were asked to connect their roller coasters to prior mathematics learning. This is important because it is not always evident how mathematics is related to a project, and this became obvious when another teacher at the school asked, "how is this math?" As an example, at Six Flags they estimated the height of a roller coaster using a clinometer and found the scale factor of their model roller coaster. They used their scale factor to find the length, width, and distance of the track if their own roller coaster were enlarged to be big enough to be a full-sized roller coaster. The calculations reveal participants' quantitative reasoning when no values were given to them. Imagining that their model was scaled large enough to be a roller coaster at Six Flags engaged participants in thinking about whether their values made sense as a life-sized roller coaster. Perhaps without realizing it, all groups learned to reason abstractly as they imagined what would be "scary" for a marble, though marbles do not experience fear. Further, none of the participants complained that their models do not actually fit the definition of a roller coaster: "an amusement park attraction that consists of a light railroad track with many tight turns and steep slopes, on which people ride in small fast open cars" (Dictionary.com). In fact, their models did not have railroad track or cars, but they did have funnels and drops, which are more common on water slides than coasters.

Construct Viable Arguments and Critique the Reasoning of Others

Participants construct viable arguments and critique reasoning when working together on any group project (Cohen & Lotan, 2014). It was particularly important in this project because if something was not quite right on the roller coaster, it simply did not work. As an ill-structured task (Silver, 1994), there are infinite ways to approach it, place specific components on the roller coaster, and problem pose. Because of this, there was a lot of trial and error that took place during construction. For example, *Orange Group* constructed their frame with a horizontal orientation (i.e.; the original frame was wide but not tall). Once they finished constructing the frame, they agreed to turn it vertically so they would have more room for vertical drops. Their idea was that if they had more vertical space, they could build up more speed where it was needed. Students also came up with their arguments as to why they should win specific awards during the presentation. The *Blue Group* argued that their group should win the Best Engineered award because they used thick, sturdy paper materials such as paper bowls for turns and paper towel rolls for beams. All groups argued as a part of their presentation why they had the best designs for specific awards.

Model with Mathematics

Unlike "modeling" found in textbooks, we used the following definition "...model-eliciting activities usually involve mathematizing – by quantifying, dimensionalizing, coordinating, categorizing, algebratizing, and systematizing relevant objects, relationships, actions, patterns, and regularities" (Lesh & Doerr, 2003, p. 5). We feel that modeling does not look like a word problem where everything is given that is needed to solve the problem. Instead, modeling in mathematics is when a question is posed, and the student must figure out what is needed to solve the problem and use what they learned to find one or more solutions (Silver, 1994; Verschaffel, Greer, & de Corte, 2002). No numbers, parameters, or size constraints were given to students. Participants chose their units of measurement when they estimated the height of the roller coaster at Six Flags and when they measured the pieces of their roller coaster. The different approaches by groups described above as well as the varied creativity in designs are evidence that this modeling project had many possibilities for final products.

Use Appropriate Tools Strategically

Participants employed many tools to complete the roller coasters. There were restrictions on what materials could be used which included paper, tape, and cardboard. There were no restrictions on tools that could be used constructing them. Participants typically asked for tools they needed: some asked for compasses to draw circles for funnels, protractors to measure the angle of depression of the drop, rulers to make straight lines and grids, and tape measures to measure the distance of the track. They got creative with some of the materials they used that still stayed within the parameters of the allowed materials. *Blue Group* used colored duct tape for construction and decoration purposes. They used the edges of paper bowls for the turns because they were already rounded to make curved track. They also printed pictures of Pennywise, balloons, and clown gloves as decorations, as well as playing "scary music" during their demonstration. All participants followed the guidelines and used clinometers, rulers, and calculators when needed.

Attend to Precision

Participants found out early on if they were not precise with measurements, everything did not line up correctly. *Yellow Group* had collapsing beams because they were not all the same size, therefore, not fitting together well. The group forced beams to fit together which made it less sturdy. They learned they had to measure out folds on the beams and make sure the measurements are consistent and precise. Students typically made tick marks at the top and bottom of a piece of paper one inch apart. This ensured flaps would be the same width. They connected the tick marks on opposite sides of the paper with a line, using five flaps per beam. Everything fit together perfectly if measured precisely. *Orange Group* found the height of the edges of tracks were very important to make tall enough so the marble would not fly out. By adding higher walls, the marble was contained and continued to work its way through the rest of the coaster. Participants quickly learned the importance of precision so that parts would work as intended.

Look for and Make Use of Structure

"Mathematically proficient students look closely to discern a pattern or structure...They recognize the significance of an existing line in a geometric figure and can use [it]for solving problems. They also can step back for an overview and shift perspective" (The Charles A. Dana Center, 2017). Participants were asked to measure the angle of depression of the drop with respect to the horizon. In order to do this, they had to place the zero degree of the protractor in line with the horizon. Though they could not draw an auxiliary line in midair, they thought about placing the zero-degree tick mark of the protractor parallel to the floor to represent the horizon. Then they measured the angle of depression. Every group took a moment at some point to shift their perspective. If there was a piece of their roller coaster that wasn't working (e.g., the marble flew off the track, the marble stopped rolling, the marble didn't make it around the loop, etc.), the group would take a step back, think about possible reasons why it wasn't working, discuss and test solutions. *Purple Group* and *Blue Group* started with a sketch of their ideas before construction (see example in Fig. 9.1). Each group made adjustments as they progressed when the coaster did not work as expected. *Orange Group* made a loop by cutting the walls of a track into small slits. They noticed if they did this to only one wall of the track, it would make a turn. They used what they learned about making a loop to make a turn. Participants were able to take something that they already did and use similar ideas for a different structure.

Look for and Express Regularity in Repeated Reasoning

Even though this project required participants to create different components such as a loop, a turn, a funnel, and a drop, there were a lot of repeated parts throughout the roller coasters. For example, the tracks were something they could test and if it worked, they could make a lot of them at once. When they knew something worked, they would work on making more of the same pieces. A member of *Yellow Group* reflected in her packet that they acted like an assembly line when making tracks.

They made one and once they knew it worked, one person made the initial cuts, the next person folded and taped it, then the last two people would tape multiple tracks together. This was something they had never made before but became quick and efficient in doing so. Every group eventually concluded that if paper folds were crisp, the structure would be stronger. Groups used a pencil to press firmly on folds so they would fold consistently. This shows repeated reasoning because they knew that if they had crisp, strong folds in the beams, it would be strong enough to hold up the roller coaster. *Purple Group* drew grids on cardstock so they could make everything consistent. Because each piece was the same size, they never had problems with pieces fitting together, unlike initial struggles from the other three groups.

Discussion

About Language

In thinking about authentic mathematics assessment for bilingual students, we believe "[Many]...teachers have not understood the importance of validating home culture and language for the development of additional language and culture understanding..." [Further], "many of the textbooks used in second language teacher education amount only to a laundry list of strategies" (Téllez & Waxman, 2006, pp. 4–5). Our purpose in incorporating language was to give students a sense of connection to their language, culture, community, and personal interest. This is what many ELL/bilingual researchers call for in bilingual learning environments (Civil, 2016; Téllez & Waxman, 2006; Thompson & Radosavljevic, 2014). It was not our intent that students would learn specific vocabulary in a particular language as a part of the project. Instead, we hoped students would practice discourse and academic language as humans that happen to be bilingual and navigate the world in two languages.

To validate culture and language, we asked all participants to argue persuasively about their projects in two languages. All participants spoke both Arabic and English as a part of their presentation though surprisingly, very little Arabic was spoken during construction. We noted examples of code-switching, direct translations, and descriptions of different aspects of projects in each language. Some transitioned between languages seamlessly, even producing typed speech notes in Arabic, while others spoke hesitantly. We believe some students were hesitant because they had not practiced using the academic language used in geometry and engineering in Arabic. We recognize, as researchers of multilingual education, that there is a distinct difference between academic language in a specific subject area, and everyday language. Bilingual students need to develop competency in both (Bunch, 2006).

Other students were simply anxious about public speaking in any language. Figure 9.2 is an example of one note card used by a participant during her speech. Similar to Zahner and Moschkovich (2011), "we think the most relevant issue is how the use of two languages facilitates (or not) students' participation in mathematical discourse practices" (p. 41) rather than identifying and labeling varied discourse practices. Interestingly, some participants were initially hesitant to use Arabic if

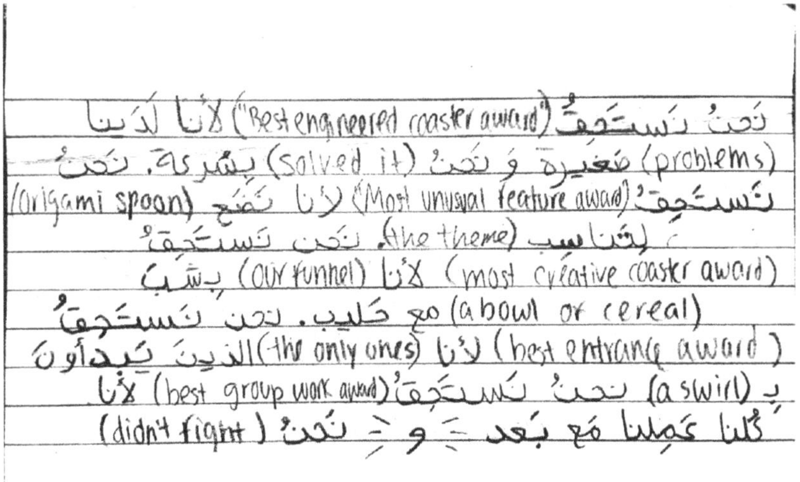

Fig. 9.2 Purple group note card

they would be required to discuss *numbers*, which some said were difficult. However, all students did present their projects using two languages, and all participants impressed the bilingual audience that subsequently voted for project awards.

One student during construction asked what to do since there was no Arabic word for "beam." In reflecting on their experience using two languages, most expressed that it was a positive experience, but some said they would not want to try it again. Here are a few sample responses:

> "I liked that we weren't being judged too hard and it was a safe place to speak it."
> "It helped me practice my more formal talk"
> "I feel good. No really, I learned that I can read Arabic in front of people."
> "I really liked that we were able to all bring out our culture and language into our school activities."

In reviewing the reflective responses which were part of the 16-page packet of mathematical work, the researchers felt they had successfully accomplished their purpose in asking students to use two languages as a part of their presentation.

About Assessment

As math teachers, it is not easy to assign a numeric score to a project when we are used to grading for one correct solution. Our advice is to create a construct map (National Academies of Sciences, Engineering, and Medicine, 2018, p. 156) that models learning progressions for all concepts related to the activity (e.g., volume, scale factor, using a clinometer, connecting to prior concepts, reasoning). Next, design a rubric to match point values and descriptions of a range of mastery for each concept. Sometimes, it may be more appropriate to assess these concepts from a competency-based perspective where a student either masters the competency, or not. For the pilot project, Miss G. presented it as a fun way to end the year. Though she assigned grades, she had not designed a comprehensive rubric for scoring. This year, she constructed a rubric to measure the concepts. Students were given the project description as well as time to review and ask questions about the rubrics that would be used to score their mastery of the concepts they were practicing. Miss G. emphasized reasoning and reflection questions. Each group had the same calculations because they worked together, but all group members had different responses to their reflection questions showing diversity in what was learned. Lots of students appear to have struggled with scale factor (taught early in the school year) but succeeded in volume (a topic taught more recently). To score a reflection question, a student received full points if they answered all parts of the question and the mathematics was correct.

Conclusion

When we set out to write about this project, we did not realize we had captured strong evidence of students learning and participating in *all eight* NCTM practice standards (The Charles A. Dana Center, 2017), in addition to creating connections to prior learning, culture, language (Téllez & Waxman, 2006; Zahner & Moschkovich, 2011), and creating a mathematics community that values errors (Boaler, 2016). While it may not be possible to always come up with a class activity that has "everything" as this seems to, we do encourage all teachers to try putting down the book, at least temporarily, and focus students on being creative and learning to value themselves and their classmates as doers of mathematics, who persevere, problematize, reason mathematically, and value the contributions of themselves and others through discourse and collaboration.

Our advice to teachers who want to try modeling with mathematics:

- Allow for enough time
- Make sure you have extra materials
- Create prior learning connections with a packet of materials
- Language, culture, and creativity help make projects personal and connected to life
- Not everyone learns and retains mathematics through text exercises; this is necessary as a part of differentiated instruction for all learners.

Finally, we wish to encourage teachers to design their own projects that support their students in hands-on learning of real-world mathematics, incorporating language, and sometimes being a facilitator of instruction instead of a lecturer/leader. Our next step is to think about how we will modify the task for a large, sometimes crowded classroom in a public-school setting.

References

Boaler, J. (2016). *Mathematical mindsets: Unleashing students' potential through creative math, inspiring messages, and innovative teaching*. San Francisco, CA: Jossey Bass.

Bunch, G. C. (2006). "Academic English" in the 7th grade: Broadening the lens, expanding access. *Journal of English for Academic Purposes, 5*, 284–301.

Civil, M. (2016). STEM learning research through a funds of knowledge lens. *Cultural Studies of Science Education, 11*(1), 41–59.

Cohen, E. G., & Lotan, R. A. (2014). *Designing groupwork: Strategies for the heterogeneous classroom* (3rd ed.). New York: Teachers College Press.

Eslami, Z., & Garver, R. (2013). English language learners and project-based learning. In R. M. Capraro, M. M. Capraro, & J. R. Morgan (Eds.), *STEM project-based learning: An integrated science, technology, engineering, and mathematics (STEM) approach* (pp. 119–128). Rotterdam, Netherlands: Sense.

Lesh, R., & Doerr, H. (2003). Foundations of a models and modeling perspective on mathematics teaching. In R. Lesh & H. Doerr (Eds.), *Beyond constructivism: Models and modeling perspectives on mathematics problem solving, learning, and teaching*. Mahwah, NJ: Lawrence Erlbaum Associates.

National Academies of Sciences, Engineering, and Medicine. (2018). *How people learn II: Learners, contexts, and cultures*. Washington, DC: National Academies Press.

Roller Coaster. (n.d.). *Dictionary.com*. Retrieved from https://www.dictionary.com/browse/rollercoaster.

Silver, E. A. (1994). On mathematical problem posing. *For the Learning of Mathematics, 14*(1), 19–28.

Téllez, K., & Waxman, H. C. (Eds.). (2006). *Preparing quality educators for English language learners: Research, policies, and practices*. Mahwah, NJ: Lawrence Erlbaum Associates.

The Charles A. Dana Center. (2017). *Standard 7: Look for & make use of structure*. Retrieved from http://www.insidemathematics.org/common-core-resources/mathematical-practice-standards/standard-7-look-for-make-use-of-structure.

Thompson, A., & Radosavljevic, A. (2014). Fostering mathematical discourse with socratic seminars. In M. Civil & E. Turner (Eds.), *Common core state standards in mathematics for English Language Learners* (pp. 81–94). Alexandria, VA: TESOL.

Tseng, K. H., Chang, C. C., Lou, S. J., & Chen, W. P. (2013). Attitudes towards science, technology, engineering and mathematics (STEM) in a project-based learning (PjBL) environment. *International Journal of Technology and Design Education, 23*(1), 87–102.

Verschaffel, L., Greer, B., & De Corte, E. (2002). Everyday knowledge and mathematical modeling of school word problems. In *Symbolizing, modeling and tool use in mathematics education* (pp. 257–276). Dordrecht: Springer.

Zahner, W., & Moschkovich, J. (2011). Bilingual students using two languages during peer mathematical discussions: ¿Qué Significa? Estudiantes Bilingües Usando Dos Idiomas en sus Discusiones Matemáticas: What does it mean? In K. Tellez, J. Moschkovich, & M. Civil (Eds.), *Latinos/as and mathematics education: Research on learning and teaching in classrooms and communities* (pp. 37–62). Charlotte, NC: Information Age Publishing.

10

What Informs Their Practice? Reflections on In-Service Teacher Education and Prior Experiences for the Teaching of Mathematics to ELLs

Amanda E. Lowry and Melinda Eichhorn

Students from all racial, ethnic, linguistic, gender, and socioeconomic groups deserve access to challenging mathematics content (National Council of Teachers of Mathematics, 2019). To make content accessible, teachers can adopt a strengths-based approach toward their students (Aguirre, Mayfield-Ingram, & Martin, 2013). In other words, students will not learn math "in spite of" or "regardless of" who they are, but they will learn math "in light of" who they are and because of the various experiences they bring to the classroom (Aguirre et al., 2013, pp. 9–10). In order to create learning environments that are accessible, challenging, and engaging for all learners, teachers can use differentiated tasks and Universal Design for Learning (UDL) guidelines (CAST, 2018; NCTM, 2019). Another practice that fosters equity and access for all students

A. E. Lowry (✉)
Rutgers University, New Brunswick, NJ, USA
e-mail: amanda.lowry@rutgers.edu

M. Eichhorn
Gordon College, Wenham, MA, USA
e-mail: melinda.eichhorn@gordon.edu

L. C. de Oliveira and M. Civil (eds.), *Teaching Mathematics to English Language Learners*, https://doi.org/10.1007/978-3-030-48355-5_10

is "making strategic use of human and materials resources" (NCTM, 2019, para. 1). Schools and districts can offer "high-quality educational opportunities" for teachers at various stages in their career to ensure that teachers have the skills, knowledge, and dispositions to support equitable and effective mathematics instruction (NCTM, 2019). However, school districts cannot just offer good professional development (PD) in order for change to take place. Teachers must participate in collaborative processes after the PD in order for change to occur (Fullan, 2001; Hargreaves & Evans, 1997), which is a shift from the commonly experienced isolation of mathematics teachers (Stein, Silver, & Smith, 1998).

Teacher content knowledge and pedagogical skill impact student learning in mathematics (Hill, Schilling, & Ball, 2008; Shechtman, Roschelle, Haertel, & Knudsen, 2010; Shulman, 1986), and in particular English Language Learners (ELLs) (Sorto, Wilson, & White, 2018). To improve the teaching and learning in content classes for ELLs, teachers should support content-specific language development, scaffold learning by building on prior knowledge (Echevarria, Vogt, & Short, 2004), and allow the use of students' first language in the classroom (Turner & Celedón-Pattichis, 2011; Turner, Domínguez, Maldonado, & Empson, 2013; Zahner & Moschkovich, 2011). However, classroom implementation of PD strategies can vary from teacher to teacher. The purpose of this chapter is to explore how two high school teachers use various experiences to inform their teaching of mathematics to ELLs in a large suburban school. Through a case study approach, these teachers reflect on their education, PD, and school supports and how those experiences have prepared them for teaching ELLs in high school mathematics classes such as Algebra 1 and Geometry. Research questions include:

- How do two teachers reflect upon and adjust their practice, based on district-wide PD (including sheltered English instruction), previous teaching experience, and self-initiated research for teaching mathematics to newcomers and English Language Learners?
- In which areas do these two teachers desire additional PD and resources?

Developing the Language of Mathematics

Developing academic language in the mathematics classroom is important, as the Common Core State Standards call for students to use language to support their mathematical practices through explanation, justification, communication, and response (Bunch, 2013). Students develop mathematical proficiency through social interactions and language. Teachers must be flexible and responsive to what emerges through these interactions in their mathematics classroom (Schleppegrell, 2010). Bunch (2013) also calls for teachers to have some background and knowledge about linguistics to help students develop language skills with their content area. He argues that "the pedagogical language knowledge of mainstream teachers can be construed as knowledge of language directly related to disciplinary teaching and learning and situated in the particular (and multiple) contexts in which teaching and learning take place" (Bunch, 2013, p. 307).

While it is imperative that teachers are aware of the language and vocabulary needed within mathematics, they should also be knowledgeable about how students learn and develop those linguistic skills. Teachers should ensure that all students are able to converse about mathematics, regardless of their language proficiencies. "The importance of language in mathematics learning illuminates the need for speaking a language that everyone is able to understand" (Meyer, Prediger, César, & Norén, 2016, p. 47). More specifically, using sheltering as a methodology specifically supports ELLs in content area classes through the use of visuals, academic scaffolding, and content-specific language that matches students' English language abilities (Crawford & Reyes, 2015). Scaffolding, which benefits all students, supports the use of sheltering for ELLs by building on what they already know, focusing on conceptual understanding, modeling content expertly, encouraging questioning, providing appropriate and frequent feedback, and creating opportunities for participation (Crawford & Reyes, 2015).

Many school districts have implemented training in sheltering and scaffolding for their teachers through the Sheltered Instruction Observation Protocol (SIOP), a highly structured framework that guides teachers through the planning and implementation of lessons to the assessment and reflection on lessons designed to teach ELLs. It is intended to be beneficial across all subject areas and to support language development alongside content area mastery. While SIOP espouses to provide ELLs with sheltering and scaffolding, it does so in a highly structured and prescriptive model, and has been highly criticized for not being "scientifically validated" by rigorous studies and encouraging passive rather than active learning (Crawford & Reyes, 2015, p. 58). While this chapter in no way endorses the SIOP model or associated PD, it is promoted in schools across the United States, including the high school associated with this case study.

Professional Development in Mathematics Education

PD is a "learning experience" which can consist of university courses, on-site assistance, action research, study groups, coaching and mentoring, examining student work and thinking, professional networks, distance education, and in-service workshops (Clements, 2004; Guskey, 2000, p. 121). PD should take into consideration each teachers' background, experience, and current context/role (Clements, 2004). All teachers are not the same; they have differing ages, stages of career, life experiences, gender, and motivation (Collinson, 1994). However, many instances of PD are homogenous: "centrally planned, uniformly provided, and regulated within hierarchically organized structures" (Sykes, 1999, p. 155). Instead, PD should be flexible and differentiated, similar to the learning opportunities we expect teachers to give to students. There is no one-size-fits-all model to learning. Teacher PD should be designed to meet the professional and personal needs of individual teachers (Day, 1997). Overall, we should teach teachers as we expect them to teach students.

PD programs are most effective when they involve the teacher and provide high-quality supervision (Clements, 2004). Supervision can include observations, experimentation, and mentoring (and allow time for reflection). Also, PD should be relevant, practical, and address teachers' needs and concerns. Math PD should "address mathematics knowledge as well as mathematics education" (Clements, 2004, p. 65).

In other words, teachers must know mathematical facts, concepts, and procedures, but also understand the relationships between and among them. Additionally, teachers must know how to present these mathematical ideas to students. Teachers should also know how children's mathematical thinking and knowledge develop. Math PD is most effective when it is grounded in a research-based curriculum and students' learning trajectories (Clements, 2004). PD should also take into consideration how teachers learn best, or adult learning styles (Clements, 2004). PD must make a meaningful connection between teachers' aspirations, the needs of the school, and the PD topic/objectives (Mack, 2000).

PD involves deeper aspects of teachers' general pedagogical knowledge, including teachers' *beliefs* about "learners, how they learn, and how that learning can be fostered by teaching" (Borko & Putman, 1995, p. 39). Teachers' knowledge of strategies and procedures is intertwined with their knowledge and beliefs about how children learn and think, and how teachers can facilitate their students' learning (Borko & Putnam, 1995).

Sustainable and long-term change in teachers' beliefs takes time and involves a continual and iterative process of growth (Guskey, 2000; Ralabate & Nelson, 2017). According to Guskey (2000), if teachers make changes to their classroom practices and see evidence of student learning, then there is a significant change in teachers' attitudes and beliefs. Usually, this change occurs after teachers see evidence of improvements in student learning, not after a PD session or course (see Fig. 10.1; Guskey, 2000).

Expert teachers maintain high expectations and consistently reinforce the expectation that all students can meet high standards for student work, effort, and behavior through effective effort, rather than innate ability (Massachusetts Department of Elementary and Secondary Education, 2018; Ralabate & Nelson, 2017). We lower our expectations of

Fig. 10.1 Model of teacher change (Guskey, 2000, p. 139)

culturally and linguistically diverse students when we perceive that their lives (academic, home, social, and emotional) will be too challenging to complete rigorous mathematical tasks (Ralabate & Nelson, 2017). Teachers can maintain high expectations for ELs by exposing them to rigorous academic vocabulary and content, while providing scaffolds and support for comprehension and offering flexible ways and choice for students to express what they know. Maintaining the same learning goal for all students, while offering flexibility and scaffolding (rubrics, online translations, checklists, etc.) within tasks, allows all students to meet high expectations (Ralabate & Nelson, 2017).

Reflective practice happens when a teacher regularly reflects on the effectiveness of lessons, units, and interactions with students, both individually and with colleagues, and uses insights gained to improve practice and student learning (Massachusetts Department of Elementary and Secondary Education, 2018). Teachers may notice little or no growth, which can be frustrating. Regardless, teachers must choose to hold high expectations of their students over time and offer support to students to access the challenging content.

Methodology

This study employs a descriptive case study approach (see Wiebe, Durepos, & Mills, 2010) in order to describe how education and PD is received and then used by two high school English as a Second Language (ESL) math teachers. Both teachers were expected to provide mathematics instruction to students in the school's ESL program. Participants' classes were specifically designated as ESL math courses and were self-contained math classes for ELLs of any level of proficiency in English who needed that math course (i.e., Pre-Algebra, Algebra 1, Geometry). The curriculum for these courses aligned with the school's general education math courses, but it was generally accepted that the pacing might be slower, that these classes might not complete the full curriculum, and that the math teachers should support language development within the context of the curriculum. As a result, some of these classes do not complete the math curriculum, which raises concerns about students' exposure to the curriculum, but that is outside the scope of this study. These two case studies were chosen because they capture differing circumstances of both monolingual and bilingual teachers. We do not assume that the two participants and their experiences are generalizable, but they do shed some light on how teachers are prepared for working with ESL adolescents in the math classroom. These two participants were purposely selected from the same high school to show the differences in teacher preparation and training within the same teaching environment in order to eliminate any impact a particular school or district may have on their teaching practice.

As researchers we considered the various areas in which teachers learn pedagogical and methodological strategies, including training, both as pre-service teachers and in-service professionals (Clements, 2004; Guskey, 2000). This educational model includes classroom experience, PD, undergraduate and graduate education, and other training avenues. Each of these different learning venues for educators and their reflection on practice, influence the day-to-day activities in the classroom (Massachusetts Department of Elementary and Secondary Education, 2018). Figure 10.2 shows the interaction between each of these aspects of professional growth and development. Teachers' background, in terms of previous training and prior teaching experience, influences how they approach teaching and learning. This also impacts how they reflect on their own practice through analyzing data, examining issues, and developing new approaches to improve teaching and learning. While all teachers in any given school may attend the same PD session, for example, their implementation of the strategies provided can vary based on their personal background and training. This fundamental idea shaped how we looked at the issues raised by each of the participants in this study, specifically in the context of high school mathematics teachers.

Data Collection and Analysis

The data collected included one-one-one interviews with each teacher, reflections on teaching practices, and a list of school-sponsored PD. The interviews and reflections were collected between November and

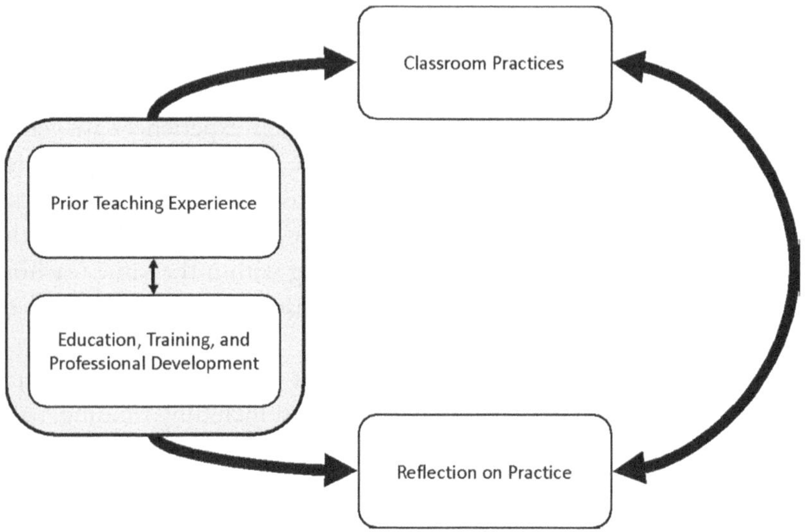

Fig. 10.2 Framework for experiences informing the teaching of math to ELLs

December of the 2018–2019 school year. Each participant participated in one interview, which was audio-recorded and transcribed. Once transcription was completed, each participant was sent a copy of the transcript for verification of information and asked if they wanted to include anything else. Neither participant added any additional information nor requested a change to their transcript.

Once transcripts were verified, the authors coded for the following: reflection on practice, obstacles, prior teaching experience, PD, adjustments to practice, resources needed, self-initiated research, strategies and methodologies in the classroom, and student strengths. We each coded the transcripts individually and then compared our coding with one another to ensure alignment. From the qualitative analysis, we were able to determine overarching themes and ideas about how PD and teacher education has prepared these teachers to work with this unique population.

Case Studies

The teachers who participated in this study both worked at West Central High School (pseudonym). The school is situated in a relatively high socioeconomic, suburban area on the East coast and the students are predominantly white. The school has a reputation for college preparation and sends students to competitive and selective universities each year. West Central High School averages about 3200 students each school year, with approximately 2.5% of the school population is identified as ELLs. This population is transient, with students enrolling and withdrawing throughout the school year. Many are students with limited, interrupted, or inconsistent formal education (SLIFE), and their educational needs differ as they must be supported both in academic content, language, and classroom behaviors (DeCapua & Marshall, 2010). The majority of the ELLs at West Central High School are from Central and South America. Approximately 80% of the school's ELL population speaks Spanish, but there is a total of ten languages represented among the students.

Generally, the school provides resources for the ESL program in terms of books and supplies. They do have varying levels of mathematics instruction, ranging from Foundations of Math to Geometry, that are offered in an ESL format in order to minimize the linguistic demands in the math classroom. In order to meet the goals of these courses, teachers must be familiar with best practices for teaching adolescent ELLs in the math classroom. The two teachers in this study have varying levels of experience and training for working with this population, as is shown in Table 10.1.

Ms. Adams

Ms. Adams, a Latina woman in her mid- to late thirties, was herself once an ESL student after her family immigrated to the United States from Mexico when she was thirteen years old. She remembers the struggles she faced in learning high school math alongside English, although she acknowledges that her experience differs from many of her students,

Table 10.1 Participants teaching experience

Pseudonym	Current position	Years in current position	ESL math courses taught	Prior teaching experience
Ms. Adams	Mathematics teacher	2	Foundations of Math Pre-Algebra Algebra 1	8 years in private schools across the country
Mr. Brown	Mathematics teacher	3	Geometry	2 years in public HS—Special Ed

since she had a "solid" and uninterrupted education in Mexico. Due to her bilingual skills, she has supported her Spanish-speaking students in their first language when the entire class spoke Spanish, but this was not her original plan when she became a teacher. Ms. Adams was always good at math and pursued both a bachelor's degree and master's degree in mathematics. It was unclear what she would do with her education until she settled on teaching during her graduate studies. Her knowledge of Spanish alongside her mathematics content knowledge allows her to make mathematics more "socioculturally and linguistically meaningful for [her students]" (García, Arias, Harris Murri, & Serna, 2010, p. 138). Ms. Adams notes that ELL "students bring real life experiences to the math classroom. They get a better grasp of concepts when they are related to real life applications. Additionally, they bring diversity to the math class since they have learned basic math skills through different methods."

Prior to her current teaching position in a public school, she taught at private boarding schools in Texas and Connecticut. When she was hired to teach ESL math, she was unprepared to meet the needs of the students in her classes when working with this specific population. She supports students using both Spanish and English, even though her courses were not explicitly bilingual. She generally used Spanish to provide additional support and clarification in a one-to-one manner for students who speak Spanish. For non-Spanish speakers, she used English for individualized instruction. In addition to the linguistic diversity in her classes, Ms. Adams shared, "the biggest shock for me was the behavior part. That's, that can be draining some days... I don't think it was prepared for that..." Ms. Adams provides "emotional support" and helps students learn "how to navigate the system so that they get the help they need, emotionally and academically." Some students are "so far behind that they're feeling defeated, so I'm having a hard time bringing them in and helping them find ways to catch up and move on." Following PD at West Central High School, Ms. Adams uses more visuals and activities embedded in their personal experience (doing yard work and earning an hourly wage), since the training made her more aware of the difficulties and overall exhaustion students feel as they try to learn a new language and understand math concepts.

Additionally, she had never been through a formal teacher preparation program, as it was not a requirement at the private schools where she worked, and thus participated in the state's alternate route to certification program. This certification process required that she be highly qualified in the subject area by college courses and then included classes provided through a local university on methodology for teaching. While this program requires 200 hours of seat time in courses that cover methodology, it does not require a practicum because you must have a teaching job prior to enrolling in this program. Since she had received both a bachelor's degree and master's degree in mathematics, she had enough content knowledge for certification, and her prior teaching experience contributed to her participation in the alternate route program.

Mr. Brown

Mr. Brown, a white male in his mid- to late twenties, attended an undergraduate program in education with the intention of becoming a high school math teacher. His program focused on content courses and general methodology classes, but Mr. Brown does not feel that he was well prepared to work with diverse groups of students. He reflected on the one course he took that touched on basic strategies for working with students with different learning needs, including students with learning difficulties and ELLs. "I only had one [course] I had to take. I believe it was called Exceptional Learners, which was just for anyone who was of special needs. It covered a very wide range of special needs students along the entire spectrum. But nothing that actually focused exclusively on just English language learners." Beyond that one course, he had no other instruction or training for working with diverse students, including ELLs, in his teacher education program.

When Mr. Brown started teaching at West Central High School, he had already completed two years of high school teaching at another public school. He was hired at his first school to teach math, but the school faced a teacher shortage and he was assigned to teach special education math classes. This posed a challenge for Mr. Brown because he had limited training in working with students with learning differences. He struggled in teaching these classes but worked hard to do the best he could. Eventually, he was let go because he did not have the appropriate certification to teach the special education math courses.

It was at this time that he was hired at West Central High School. Again, Mr. Brown was hired as a math teacher, but was asked to teach the ESL Geometry course. This would not have been his first choice in teaching assignment as he had limited training and experience with ELLs, but he also knew that there was little space to express concern about the assignment. He had hoped that it would be a one-year assignment, but it has since turned into, at the time of this study, a three-year teaching assignment. The only ESL math course that Mr. Brown taught was the ESL Geometry course, which typically was for students in their third year of high school and English acquisition courses. Occasionally, students in this course transfer in high school credits form their prior schools and take ESL Geometry in their first year in the United States, but that is not a regular occurrence.

Mr. Brown expressed frustration with the training provided by the school, as it was too similar from year to year and didn't provide enough math content-specific strategies. Mr. Brown acknowledges that language is a barrier for him, since he speaks only English. He struggles when students shut down, and he cannot communicate with them in their first language. However, since Geometry is quite visual, he can provide students with visual supports, but adds that "if students are also struggling with other areas in terms of problem solving, their algebraic skills…

vocabulary… it's only going to make Geometry that much more challenging." Mr. Brown stated that classes which have in-class support "have always performed best." In one of his ESL Geometry classes, there is a bilingual paraprofessional in the room, which has helped students gain access to math concepts when the content is translated by the para. Mr. Brown also uses guided notes. However, Mr. Brown did not gain teaching strategies for his ESL Geometry course from the school-based PD, since the training "was more so a smattering of like ideas, techniques all across the board. And I feel like when you do things as contents all across the board oftentimes math is the first one to get left out. And I feel like not enough was focused on that."

Overarching Themes

While Ms. Adams and Mr. Brown had different paths to their current roles as ESL math teachers at West Central High School, many of their teaching experiences working with this population are similar. They have received similar in-school PD, feel unprepared to meet the socio-emotional as well as the mathematical needs of the students in their classes, and would like additional PD as they continue in their roles as ESL math teachers. Additionally, they have tried different strategies building upon their personal skill sets to varying levels of success for their individual classrooms and have adapted their teaching practices over time.

In-School Professional Development

Mr. Brown had two years of school-sponsored PD, while Ms. Adams only had one due to when they were hired at West Central High School. This may influence how Mr. Brown feels about the success and impact of the trainings, as he felt the second year was repetitive of the first year. Table 10.2 shows the number and type of trainings on working with ELLs offered by West Central High School for teachers. Teachers must

Table 10.2 List of PD with a focus on ELLs offered by West Central High School over a three-year period

Training name	When offered	Offered to?	Ms. Adams' and Mr. Brown's participation?	Content
Summer PD	Each summer—a full or half day at school	ESL and Bilingual teachers (mandatory) ESL content teachers (voluntary)	Neither	Changes each summer. Topics include assets-based teaching, overview of bilingual programs, etc.
New teacher orientation	Each summer—30 minutes allocated during a multi-day orientation	New teachers to the school district	Ms. Adams and Mr. Brown	Basic overview on ELL population and general modifications
Sheltered instruction training	Each school year—3 full day trainings	One cohort per year of content area teachers. Those with ESL students have priority	Mr. Brown—2 years Ms. Adams—1 year	
One-hour PD	School in-service day. Once per year	Teachers who have interest	Neither	Basic overview (1 hour)

complete twenty hours of PD each school year to meet state requirements and the opportunities listed in Table 10.2 count toward those PD hours.

In reflecting on the PD he received from the school, Mr. Brown felt discouraged during the first day of the three-day training in the second year since it was "very similar" to the training from the previous year. He continued by saying, "If you're going to do it and provide more training and more techniques and whatnot. But I felt as though that they were too similar and we're not as helpful as they could have been." Ms. Adams also reflected on the PD trainings offered and expressed a different take away from them. As a former ESL student herself, she was able to view the trainings from the eyes of the students. "They did a good job of letting us see how it feels to be an ESL student, like how difficult it is trying to make sense of what the teacher is saying when it's in a different language."

While their reactions to the school provided trainings differed, both teachers felt that the PD would have been more helpful if the content was more math specific. Both wish that there was more math-specific training on how to best work with ELLs, but neither have explored any outside PD on their own, primarily due to time constraints. They do not feel confident that they have received training to best teach math to their students, even though they are both confident in their abilities as math teachers. The school-sponsored PD would have been more beneficial to them if they had received specific strategies to improve math instruction for this population.

Realities in the Classroom

As is shown in Fig. 10.1, teachers receive PD but then have to implement the takeaways to their classroom. The strategies provided through PD are implemented in varying degrees due to external supports and stressors that are not considered in the PD. Both Ms. Adams and Mr. Brown described some of these outside forces that impact their classrooms that were not addressed in their PD sessions. These situations affected their teaching both positively and negatively.

Strengths-Based Approach

Ms. Adams views the diverse backgrounds in her classroom as an asset and something that should be valued and developed:

> ESL students bring real life experiences to the math classroom. They get a better grasp of concepts when they are related to real life applications. Additionally, they bring diversity to the math class since they have learned

> basic math skills through different methods. Also, and while this is not necessarily a strength in math, but rather socially, these students show a great deal of empathy towards their classmates and most of them look out for each other. (personal communication, November 21, 2018)

As Ms. Adams describes, her students have valuable ways to contribute to her classroom environment. It is important that teachers recognize their students' skills, backgrounds, and experiences socially, linguistically, and mathematically in order to support their sensemaking and development as students (Aguirre et al., 2013).

Small Class Sizes

The ESL math classes at West Central High School are generally capped at fifteen to eighteen students and frequently have only ten students in them at the beginning of the year to leave space for new students to enroll. Mr. Brown's ESL Geometry classes have had enrollments as low as three students due to students withdrawing from school. While a smaller class size could be better for both students and teachers, Mr. Brown found it difficult at times due to high levels of absenteeism even for the few students left in his classes. Ms. Adams also commented on the transient nature of the students and high levels of absenteeism as challenges for keeping all students current with the math content in her classes. She mentioned that many of the students work long hours and have little time for sleep and homework. If they miss their bus, they may not have any way to get to school.

Language as a Support or Barrier?

Both Mr. Brown and Ms. Adams mentioned language as a challenge, but in different ways. Mr. Brown mentioned that he has no second language skills and must rely solely on English to engage his students. Students then get frustrated if they don't understand him, which lends itself to disengagement in the Geometry lesson. It is important to note that one of Mr. Brown's sections of ESL Geometry had a bilingual paraprofessional and the other did not. This caused some tension between his two classes because one class felt as though they were not given the same linguistic support as the others. Ms. Adams, while bilingual, mentioned that there are many instances when she is trying to support students in Spanish, but she does not know the academic language in Spanish since she studied math in English as a high school student once she immigrated to the United States. Regardless of their Spanish language skills, both teachers felt ill-prepared for helping bridge the divide between academic and everyday language in both English and Spanish. Both Mr. Brown and Ms. Adams agree that having an additional person in the classroom is beneficial for helping with language and content.

Social Emotional Needs

Since many of West Central High School's ESL students have survived traumatic experiences, both teachers are concerned about how to best support this population's emotional and mental health needs. The aftereffects of these experiences sometimes result in outbursts, defiance, and foul language in the classroom. Also, some of these same students have had limited exposure to education and the teachers are expected to teach them how to be a student, while supporting students' emotional and mental well-being. These are not defining characteristics of all ELLs, but for those that this is true, both Mr. Brown and Ms. Adams are not sure how best to support them while delivering academic content. While the school does have two bilingual guidance counselors whose role is primarily academic as opposed to addressing individual student's emotional issues.

Adjustment to Practice

In light of these circumstances and as a result of their PD experience, both teachers have adapted their practice to try to meet these needs. Ms. Adams specifically has tried to reword her application problems to be in terms that the students understand. Some ways that she has done this is by using "kilograms instead of pounds because that's what they're familiar with." She tries "to make connections between the math and something that they are familiar with" and that they can connect to the jobs they do after school, like yard work or hourly pay. Similarly, Mr. Brown "definitely [doesn't] do the exact same things from [his] first year." He has changed his lesson delivery as well and now uses guided notes for instruction. He commented, "I have also slowly changed the way I do give my notes. I give different packets, um, still cover the same topics, but they're a bit more structured and laid out in the guided notes." Lastly, he has reduced the amount and type of homework assignments he gives in order to reduce the out of school demands on students. As a result of the PD, these adaptations to teaching practice were made to benefit the students and their learning potential, which is an important focus for any teaching modifications for ELLs. The way Ms. Adams and Mr. Brown have responded to obstacles over time shows how their expectations are changing—just at different rates. Since this student group's needs change each year, teachers of these mathematics classes must be observant to recognize their student's strengths and needs and flexible to adapt their teaching practice to benefit their students.

Although both teachers know that they have gaps in their knowledge and skills for teaching ELLs, they seem unsure what resources are out there or what to ask for in terms of additional PD. Ms. Adams commented, "I'm not sure. I feel like I could probably use more resources or more help. [...], I'm still learning. I'm not too sure what it is that I need." This, along with their busy schedules and desires to teach more than just ESL math, makes it difficult for Ms. Adams and Mr. Brown to prioritize furthering their professional toolkit for teaching math to ELLs. Both teachers feel that their teacher training programs provided limited instruction on how best to teach math to ELLs and feel that there should be more math-specific strategies for working with this population. Additionally, they would like to have more time to focus on furthering their skill set in this area.

Recommendations

Teacher education programs and school districts can help alleviate the challenges math teachers of ELLs face by providing the PD and instruction previously mentioned as lacking for Ms. Adams and Mr. Brown. Ms. Adams would specifically like more PD on how to support students in learning math while being supportive of their emotional and mental health needs, while Mr. Brown would like PD on making math content more accessible to students when there is a language barrier. More specific training for math teachers on how to support language acquisition for high school ELLs while addressing the mathematics content would be key. It would also be beneficial if teacher preparation programs taught teachers how to modify their assignments and word problems to be less linguistically demanding while incorporating language and context that ELLs are familiar with. The training and PD must be ongoing and continuous (Guskey, 2000; Ralabate & Nelson, 2017), but, as mentioned by both participants, also relevant and specific to the math classroom (Clements, 2004). This is not to say that more general sheltered instruction training was not beneficial for Ms. Adams or Mr. Brown, but they now feel that more math-specific training would improve their practice.

Teachers need to tailor their general best teaching practices to the needs of ELLs, even though there is some overlap between teaching strategies that benefit all students—including English Learners as well as students in special education (see Lowry, Eichhorn, & Burke, 2019). As teachers meet the needs of ELLs, all students will benefit (Oliver, 2017). See Eichhorn, Lowry, and Burke (2019) or Lowry et al. (2019) for some examples of strategies that promote mathematics learning for adolescent ELLs. Other sheltering and scaffolding strategies for mathematics include: board games, learning centers, response journals, student-generated word problems, think alouds, and think-pair-shares (Crawford & Reyes, 2015). Additionally, there are reasonably priced online courses that focus on teaching ELLs in high school mathematics classes offered through TESOL (see: *Self-Study: ESL for the Secondary Mathematics Teacher*).

References

Aguirre, J., Mayfield-Ingram, K., & Martin, D. (2013). *The impact of identity in K-8 mathematics: Rethinking equity-based practices*. Reston, VA: NCTM.

Borko, H., & Putman, R. T. (1995). Expanding a teacher's knowledge base: A cognitive psychological perspective on professional development. In T. R. Guskey & M. Huberman (Eds.), *Professional development in education: New paradigms and practices* (pp. 35–65). New York: Teachers College Press.

Bunch, G. C. (2013). Pedagogical language knowledge: Preparing mainstream teachers for English learners in the new standards era. *Review of Research in Education, 37,* 298–341.

CAST. (2018). *Universal Design for Learning Guidelines version 2.2*. Retrieved from http://udlguidelines.cast.org.

Clements, D. H. (2004). Major themes and recommendations. In D. H. Clements, J. Sarama, & A. M. DiBiase (Eds.), *Engaging young children in mathematics: Standards for early childhood mathematics education* (pp. 7–72). Mahwah, NJ: Lawrence Erlbaum Associates, Publishers.

Crawford, J., & Reyes, S. A. (2015). *The trouble with SIOP: How a behaviorist framework, flawed research, and clever marketing have come to define—and diminish—sheltered instruction for English language learners*. Portland, OR: Institute for Language and Education Policy.

Collinson, V. (1994). *Teachers as learners: Exemplary teachers' perceptions of personal and professional renewal*. San Francisco: Austin & Winfield Publishers.

Day, C. (1997). Teachers in the twenty-first century: Time to renew the vision. In A. Hargreaves & R. Evans (Eds.), *Beyond educational reform: Bringing teachers back in* (pp. 44–61). Buckingham: Open University Press.

DeCapua, A., & Marshall, H. W. (2010). Serving ELLs with limited or interrupted education: Intervention that works. *TESOL Journal, 1*(1), 49–70.

Echevarria, J., Vogt, M., & Short, D. (2004). *Making content comprehensible for English learners: The SIOP model* (2nd ed.). Boston, MA: Pearson.

Eichhorn, M., Lowry, A. E., & Burke, K. (2019). Increasing engagement of English learners through universal design for learning. *Journal of Educational Research and Practice, 9*(1), 1–10.

Fullan, M. (2001). *The new meaning of educational change* (3rd ed.). New York: Teachers College Press.

García, E., Arias, M., Harris Murri, N. J., & Serna, C. (2010). Developing responsive teachers: A challenge for a demographic reality. *Journal of Teacher Education, 61*(1–2), 132–142.

Guskey, T. R. (2000). *Evaluating professional development*. Thousand Oaks, CA: Corwin Press.

Hargreaves, A., & Evans, R. (1997). Teachers and educational reform. In A. Hargreaves & R. Evans (Eds.), *Beyond educational reform: Bringing teachers back in* (pp. 1–18). Buckingham: Open University Press.

Hill, H., Schilling, S., & Ball, D. L. (2008). Unpacking pedagogical content knowledge: Conceptualizing and measuring teachers' topic-specific knowledge of students. *Journal for Research in Mathematics Education, 39*(4), 372–400.

Lowry, A. E., Eichhorn, M., & Burke, K. (2019). Why recreate the wheel when collaboration is key? Implementing strategies for success across student group. *Teaching for Equity and Excellence in Mathematics* (TEEM), *10*(1), 6–15. Retrieved from https://toma.memberclicks.net/assets/documents/TEEM/TEEM10FINAL.pdf.

Mack, P. (2000). Read two chapters and call me in the morning: The case against prescriptive staff development. In P. Senge et al. (Eds.), *Schools that learn: A fifth discipline fieldbook for educators, parents, and everyone who cares about education* (pp. 381–383). New York: Doubleday.

Massachusetts Department of Elementary and Secondary Education. (2018). *Massachusetts model system for educator evaluation: Classroom teacher rubric.* Retrieved from http://www.doe.mass.edu/edeval/model/PartIII_AppxC.pdf.

Meyer, M., Prediger, S., César, M., & Norén, E. (2016). Making use of multiple (non-shared) first languages: State of and need for research and development in the European language context. In *Mathematics education and language diversity* (pp. 47–66). Cham: Springer International Publishing.

National Council of Teachers of Mathematics (NCTM). (2019). *Access and equity in mathematics education.* Retrieved from https://www.nctm.org/Standards-and-Positions/Position-Statements/Access-and-Equity-in-Mathematics-Education/.

Oliver, G. (2017). Supporting English language learners in secondary science through culturally responsive teaching. In L. C. de Oliveira & K. Campbell Wilcox (Eds.), *Teaching science to English language learners: Preparing pre-service and in-service teachers* (pp. 111–136). London: Palgrave Macmillan.

Ralabate, P. K., & Nelson, L. L. (2017). *Culturally responsive design for English learners: The UDL approach.* Wakefield, MA: CAST Professional Publishing.

Schleppegrell, M. J. (2010). Language in mathematics teaching and learning: A research review. In *Language and mathematics education: Multiple perspectives and directions for research* (pp. 73–112). Charlotte, NC: Information Age Publishing.

Shechtman, N., Roschelle, J., Haertel, G., & Knudsen, J. (2010). Investigating links from teacher knowledge, to classroom practice, to student learning in the instructional system of the middle-school mathematics classroom. *Cognition and Instruction, 28*(3), 317–359.

Shulman, L. S. (1986). Those who understand: Knowledge growth in teaching. *Educational Researcher, 15*(2), 4–14.

Sorto, M. A., Wilson, A. T., & White, A. (2018). Teacher knowledge and teaching practices in linguistically diverse classrooms. In *Language and communication in mathematics education* (pp. 219–231). Cham: Springer.

Stein, M. K., Silver, E. A., & Smith, M. S. (1998). Mathematics reform and teacher development: A community of practice perspective. In J. G. Greeno & S. V. Goldman (Eds.), *Thinking practices in mathematics and science learning* (pp. 17–52). Mahwah, NJ: Lawrence Erlbaum Associates, Publishers.

Sykes, G. (1999). Teacher and student learning: Strengthening their connection. In L. Darling-Hammond & G. Sykes (Eds.), *Teaching as the learning profession: Handbook of policy and practice* (pp. 151–179). San Francisco: Jossey-Bass Publishers.

Turner, E., Domínguez, H., Maldonado, L., & Empson, S. (2013). English learners' participation in mathematical discussions: Shifting positionings, dynamic identities. *Journal for Research in Mathematics Education, 44*(1), 199–234.

Turner, E. E., & Celedón-Pattichis, S. (2011). Mathematical problem solving among Latina/o kindergartners: An analysis of opportunities to learn. *Journal of Latinos and Education, 10*(2), 146–169.

Wiebe, E., Durepos, G., & Mills, A. J. (2010). *Encyclopedia of case study research.* Los Angeles, CA: Sage.

Zahner, W., & Moschkovich, J. N. (2011). Bilingual students using two languages during peer mathematics discussions: ¿Qué significa? Estudiantes bilingües usando dos idomas en sus discusiones matemáticas: What does it mean? *Latinos/as and mathematics education: Research on learning and teaching in classrooms and communities* (pp. 37–62). Charlotte, NC: Information Age Publishing.

11

Shifting the Mathematics Learning Culture for ELL Students at "The Most Diverse High School in the United States"

Susan Holloway and Gerardo De La Garza

This case study of mathematics teachers' work takes place within a large urban public high school in the western United States that primarily serves English Language Learners (ELLs). In response to the long-term under-performance of the school, the entire school, especially the mathematics department, changed radically in order to more effectively address the needs of students. Nearly 80% of the school's students speak a home language other than English, so utilizing learning experiences that focus on mathematics and language have not been an aside, but rather the central theme of a profound reconstruction encompassing attitudes, resources, and practice.

At this high school, the recent past indicators of mathematic achievement were alarmingly low and continuing to decrease. With a proficiency rate in mathematics of less than 10%, the school was clearly failing to

S. Holloway (✉) · G. De La Garza
Aurora Central High School, Aurora, CO, USA
e-mail: swholloway@aurorak12.org

G. De La Garza
e-mail: gadelagarza@aurorak12.org

L. C. de Oliveira and M. Civil (eds.), *Teaching Mathematics to English Language Learners*, https://doi.org/10.1007/978-3-030-48355-5_11

prepare its students for future success. As a result, mathematics instruction at this school became one of the focal points in a school turnaround mandated by the state Department of Education (DoE). The path toward school turnaround is known to be notoriously difficult (Hansen & Choi, 2011; Stuit, 2010), and within the small set of successful turnaround schools, success at the secondary level is even more problematic (Smarick, 2010). Even so, the variables that contribute to success are slowly being teased out of stories from schools on this path, and the protagonist school of this story aims to implement those variables with focused intensity.

In a process directed by the state's DoE, the school's constituency of parents, students, school leadership, and community members chose to make deep foundational changes to many aspects of the learning landscape. These changes are outlined in a document named the School Innovation Application (SIA) and submitted to the state DoE (De La Garza, 2016). The initiative to improve the school relies on gains in teacher expertise, improvements in classroom environment, interventions for a set of off track students, and deeper connections with families and the larger community. Efforts within the mathematics department stem from these initiatives.

This chapter begins with a historical perspective of this High School and then describes strategies most strongly leveraged by the mathematics department to effect language learners. A following section includes the reflections of mathematics educators from the school about experiences that have most informed their teaching practice with language learners. In summarizing comments, the most current state decision related to the school are shared and remaining questions are considered.

History of School: Demographics and Surrounding Community

The work described in this chapter takes place at a large urban public high school situated in the western United States. The school was established in the early 1900s and presently serves one of the most impoverished neighborhoods in the state, where one in five residents identify as an immigrant or refugee and where the community struggles with gang issues and crime rates that exceed state averages. Recently, Chalkbeat noted that this school is the most diverse school in its home state, and speculated that it may be the most diverse high school in the United States (Robles, 2018).

Diversity and global awareness are genuine sources of pride for the school, and producing global leaders is a part of the school vision (see Fig. 11.1). However, within the past decade or so, increasing numbers of language learners and students with interrupted formal education were met by a traditional schooling system unprepared to address their needs, and this helped fuel a precipitous and downward trend in achievement scores. Mathematics proficiency fell much lower than state standards, and for ELLs it sunk even lower, placing them behind all other subgroups. In 2016, with a student body made up of nearly 80% language learners speaking over fifty different languages (J. Wos, personal communication, December 2019), nearly 90% of the entire student population fell into the category of "not proficient" in mathematics, and the school found itself within the category of the lowest 5% of mathematics achievement in the state (Colorado DoE Website, 2019).

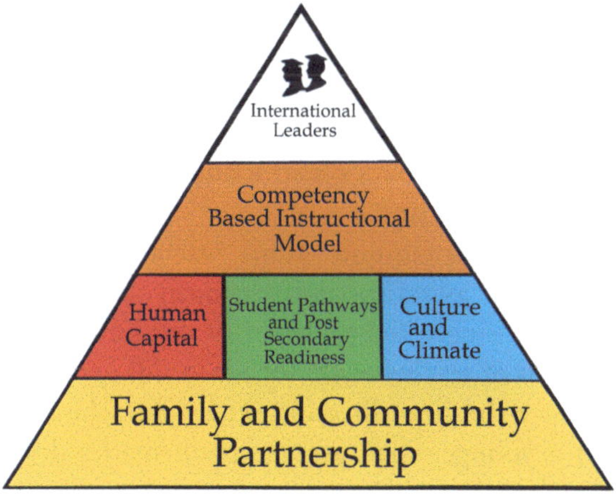

Fig. 11.1 Fostering International Leaders (Adapted from De La Garza, 2016)

As a result of this pattern of failure, the state DoE assigned the label of Priority Performance Status to the school in 2016, and the school was compelled to create a turnaround plan. This exhaustive process and analysis led to a new and profound understanding of an institution that had drifted away from, and no longer served its constituency. The resulting turnaround plan is articulated in a far-reaching document, the SIA (De La Garza, 2016). Overall, the SIA called for a deep cultural change, and the district administration provided flexibility for the new practices and programs that would support this change.

The Mathematics Department During Turnaround

The mathematics department at this school is large, made up of four full-time instructors at the 9th-grade level, three full-time instructors at the 10th-grade level, three full-time instructors at the 11th-grade level, and three full-time instructors at the 12th-grade level. The school also employs two full-time mathematics coaches, a full-time mathematics instructor dedicated to students who have alternative schedules because of parenting or full-time employment, and four special education instructors who co-teach at all grade levels in mathematics classrooms, and a dedicated instructor overseeing students in an online credit recovery program. The total number of professionals in the mathematics department offering mathematics instruction and instructional support to students approaches twenty.

As a result of the turnaround effort, teachers in the mathematics department have been asked to simultaneously address a swirl of new initiatives. Not surprisingly, the pressures of this kind of transformation produced an exodus of teachers who for many different reasons found it difficult to embrace the new direction set by the school. Presently, the school employs a younger than average mathematics department with a median age of less than thirty years; over fifteen years younger than the national average for mathematics teachers (Data USA, n.d.). Of all mathematics instructors, only two have been at the school for more than four years, and while all mathematics teachers at the school are endorsed in the state to teach mathematics, the school employs a large share of alternative licensure teachers.

Increasing Expectations for Language Learners

This section explains how the school has worked to increase expectations for language learning students. Each initiative is a response to areas of concerns identified by the SIA and submitted to the state DoE. Concerns are listed below, and each concern is followed by a description of how the mathematics department has responded to these concerns.

SIA Highlighted Concern: The mathematics department was found to lack adequate professional development and training in order to increase expectations for ELLs (De La Garza, 2016).

Response: To address the needs of ELL students in mathematics classrooms required both a shift in mind-set along with increased teacher capacity to implement specific teaching strategies.

One profound change that was made in order to improve the experience of language learners was to integrate them more fully into regular mathematics classrooms. Before the school turnaround, supports for ELLs had been confined to segregated English Language Development courses with the result that most ELLs were assigned to mathematics classes that were below grade level. The broad integration of ELL students into appropriate mathematics classes meant that mathematics teachers began to accommodate students who were more nascent in their English skills than they had served in the past. Clearly, the mathematics department required intensive training in supporting language learners in grade-level classrooms.

The mathematics department now regularly engages in specific professional development targeting instructional strategies for ELL. Planning within grade-level teams now habitually includes strategies that support acquisition of academic mathematics language including mathematics inquiry, discussion, and disciplinary writing so that students can access grade-level mathematic skills. Teachers create lesson plans that include opportunities for students to practice specific mathematics vocabulary and they cite academic language supports such as sensory supports, graphic supports, and interactive supports in these plans. Students are instructed in how to convey their mathematical ideas verbally and in writing using academic vocabulary.

Yet another method supporting increased expectations for language learners is an expanded offering of rigorous courses in the mathematics department. Before the school turnaround, the mathematics department met the unfortunate but typical practice for low-income schools in that the curriculum abounded with remedial mathematics courses that fell outside of the traditional, academic mathematics hierarchy (NCTM, 2014; Royster, Gross, & Hochbein, 2015; Tsoi-A-Fatt Bryant, 2015). One goal for the mathematics department became to increase student participation in college-level concurrent credit mathematics courses. These courses offer college credit to our students authorized by community colleges or four-year universities and are guaranteed to transfer to any four-year public institution in the state of Colorado.

Participation in these concurrent credit college courses has increased phenomenally over the past four years. The college courses offered at our school presently include Calculus I, Calculus II, College Algebra, Trigonometry, Statistics, and a Statistics Technology Laboratory. Courses are free to students, funded by a state educational initiative (Colorado Department of Higher Education, COF, n.d.). Language learners are well represented in these courses; approximately, three-quarters of all students gaining college credit in these courses speak languages other than English at home.

Planning and Collaboration Structures

SIA Highlighted Concern: The mathematics department lacked a robust structure to plan and support effective instruction aligned to Colorado's Common Core Standards (De La Garza, 2016).
Response: The mathematics department needed to provide clear expectations to all students about skills and knowledge necessary to succeed in high school and upon graduation.

Before the turnaround effort, the principal of the school noted that the failure rate in the mathematics department was over 70% (De La Garza, Explanation to teachers during PLC, Spring, 2019). Teachers as well as students urgently needed an improved roadmap to success. The systems put into place include a new vertically aligned curriculum and well-organized Professional Learning Communities (PLC).

The mathematics department leveraged newly formed grade-level PLCs to uncover obstacles to learning and collaboratively plan improved lesson content, structure, and delivery. Now, grade-level teams function as a mutually supportive and systematic structure for improving lesson planning and delivery. A formal Data Driven Instruction (DDI) cycle, described in more detail in the next section, was developed to improve each PLC's focus on what students need. Within this structure, grade-level mathematics teams choose a lesson, assessment cycle, or learning strategy that they would like to improve upon or to investigate. Teachers investigate an instructional strategy that targets language learners and is designed to increase teacher understanding of their own practice, provide accountability, and support reflection and provide a written history for the PLC efforts as teams move through the school year.

Ineffective Assessments with Resulting Unresponsive Instruction

SIA Highlighted Concern: The mathematics department needed more complete information about what each student has learned, including formal and informal assessments in order to plan further instruction (De La Garza, 2016).

Response: The mathematics department enacted a data cycle encompassing competency-based instruction, assessment and analysis of work and a reteach cycle.

A new data cycle was adopted that provides frequent feedback for mathematics instructors about how students are doing in their classroom and an opportunity for a reteach cycle for non-proficient students. The intent of this cycle is not only to ensure that every student demonstrates proficiency in a standard before moving forward, but to also provide meaningful and flexible acceleration and recuperation opportunities. This method is implemented especially with the goal of increasing opportunities to support language learners, immigrant, refugee, or those who have experienced interrupted formal education.

To support this cycle, the mathematics department added interim exams to round out its base of information from which to glean student progress information and to support the DDI structure. Interim assessments are valuable because they closely align to standards-based instruction that occurs within quarterly and semester grading cycles, are common across content areas and grade levels, align with the instructional sequence of clearly defined grade level and content expectations, re-assess previously taught standards, and can be used to prove student proficiency at skills when other evidence is insufficient for passing to the next set of standards.

Results from the interim exams allow teachers to better understand what students have assimilated in the time they have spent in mathematics classrooms. PLCs connect the learned skills to particular lessons and use the information to both create a reteach lesson and to improve instructional techniques for future mathematics lessons.

Unresponsive and Ineffective Student Support System

SIA Highlighted Concern: The mathematics department lacked an adequate response to low performing students that are absent from school an inordinate amount of time due to a variety of factors (De La Garza, 2016).

Response: The mathematics department engaged in the following strategies with the specific intention of increasing our ability to intervene with off-track students in mathematics.

The following three interventions help the mathematics department support individual low attending or troubled students: (1) a house model wherein cross content teachers can support each other in creating solutions for specific students of concern, (2) the formation of a school-within-a-school to address the most high-needs students, and (3) credit recovery for students missing mathematics credit, but not otherwise off-track to graduate. These structures provide recuperation opportunities for off track students, increasing opportunities to support ELLs, students with interrupted schooling, and under credited students.

The house model here is a student-centered approach with a goal of promoting more personal student–student and teacher–student relationships wherein students are grouped in common classes throughout their day. Houses are not a means for tracking students, and an effort is made to balance the houses with students with a diversity of academic skills. Time is provided for mathematics instructors to meet with other teachers who have the same students in their classrooms. Then, mathematics teachers share strategies and action plans with other subject area teachers about individual students.

The second intervention, the school-within-a-school model, was formed to serve the most high-needs students. These students include young parents, gang members, homeless youth, and others who have overwhelming personal circumstances making it impossible for them to find success in a traditional school setting. In this program, students receive intensive instruction in mathematics and other core subjects. The school-within-a-school has been almost single handedly responsible for increasing our graduation numbers and decreasing our drop out numbers as its organization allows for a flexible schedule and individualized learning plan for every student.

The third intervention structure involves providing two credit recovery systems; one, a traditional online credit recovery program wherein students interact with an online program to complete credits, and two, a set of standards-based, highly engaging teacher-created classes designed to provide an alternative mathematical experience for students who are off track with mathematics credits. While the teacher-created courses are intended to help retrieve lost mathematics credits, they are also designed to pique interest in mathematics by tapping high interest mathematics applications. Some examples of these classes include the study of statistics through the card game Magic©; creating engineering drawings with a focus on slope, distances, scale, areas, volumes and surface areas; using art to express mathematic symmetries in two and three dimension; and manipulating databases through the use of spreadsheets such as Excel©.

Together, the above practices and strategies have impacted the trajectory of success in mathematics for language learners.

Reflections From School Educators

The following paragraphs draw on interviews of mathematics instructors from this school to better understand how the new systems and structures described in the first section of this chapter relate to teacher perceptions of success in serving ELLs. Since teaching is considered one of the most stressful occupations (Herman, Hickmon-Rosa, & Reinke, 2018; Yu, Wang, Zhai, Dai, & Yang, 2015), and teachers at high-poverty urban schools experience much less job satisfaction and significantly more stress than teachers in other settings (Guglielmi & Tetrow, 1998; Ouellette et al., 2018), and since high teacher stress is negatively associated with students success (Herman et al., 2018; Ouellette et al., 2018) and positively associated with increased staff turnover (Fantuzzo et al., 2012; Yang, Ge, Hu, Chi, & Wang, 2009), it is of interest to identify specific factors that these mathematics teachers name as supporting them in their work.

The questions were based on areas suggested by the Job-Demand-Control Support (JDCS) model, an empirically supported model of job stress which measures job demands, job decision autonomy, personal background and critical life events (including educational preparation), and on-the-job support (Ouellette et al., 2018; Payne & Fletcher, 1983). According to the model, jobs characterized by high demands, low decision latitude, and low social support have a higher risk of stress (Kristensen, 1995). The interviewees include five mathematics instructors, and a sixth interviewee is an administrator who created and oversees the alternative school-within-a-school program. A brief description of each interviewee follows.

Interviewee #1: This teacher is currently working toward a master's degree in linguistic development. This is her first teaching assignment.

Interviewee #2: This interviewee is a third-year instructor who began teaching mathematics with an alternative license. This school was his first assignment.
Interviewee #3: This interviewee is a second-year instructor at our school. He is an immigrant and a language learner who entered teaching through an alternative licensure program.
Interview #4: This educator is the director of the school-within-a-school program and he reflects on what makes a math teacher successful in his program.
Interview #5: This instructor received her pedagogical training through the alternative licensure program, and while she has previous teaching experience, it is her first year at this school.
Interviewee #6: This instructor is a bilingual immigrant. He has a master's degree in mathematics and completed the alternative licensure program.

The following paragraphs are organized by interview topic, and the interviewee is identified by number. Some contextual information illuminates the interviewee's quotes.

Question 1: What influenced you to work in a school with large numbers of ELL students?

Interviewee #2 relates:-

- *My experience in AmeriCorps got me interested in teaching and gave me a growing passion for education. I still believe that education is "the great equalizer", so when I heard there were positions here, I applied.*

Interviewees #3, #4, and #6 are themselves bilingual teachers of color. They relate how their personal background has been an influence in their decision to work at their present school.

Interviewee #3 says:

- *I feel that I belong in a multicultural school because I am from another country. I have something to offer the multicultural students. I understand what other language learners need for help.*

Interviewee #6 says:

- I am myself a language learner. I know how the ELLs struggle with word problems, vocabulary, and even context. For example, probability questions with cards and betting are all culturally biased. Playing cards is not a good thing, it is negative in my background. We need to pay attention to these things for our students.

Interviewee #4 describes how the diverse background of the mathematics instructor in the alternative program for off track students is key to the success of the program:

- Our math instructor is a serious part of our magic. He is an immigrant and language learner himself and he understand the experience of being an immigrant. He relates to the kids and their experience in a very personal way.

Question 2: How has your efficacy as a mathematics teacher of language learners increased by understandings gained through academics or on-the-job training?

Some interviewees identified their academic preparation and student teaching experience as being significant and positive. For example, Interviewee #1 said:

- By the time I arrived at this school, I was able to initiate group worthy tasks from my academic background and research. I had not practiced the many techniques from student teaching, but I could recognize them. I could recognize instructional strategies and how they created an engaging classroom for language learning students.

Interviewee #5 also had a positive experience saying:

- I received real and important coaching during my student teaching experience. I learned how to be an active monitor of my students and it is still my best strategy for ELL students.

Other interviewees lacked such positive experiences. Interviewee #2 relates:

- Because I entered an alternative licensure program when I was hired, I began teaching before I ever took a class in pedagogy. I had over 100 students to deal with every day. Another friend and I who did the same thing still laugh about that year.

Interviewee #6 said:

- When I first arrived in the United States, I worked as a para-professional for five years in math classrooms while I worked on my master's degree. The alternative licensure program was for the most part review since I had been already in the classroom for years.

Question 3: What are the most important job embedded supports for your teaching efficacy?

School led Professional Development was not frequently mentioned as a highly applicable support system. Instead, teachers tended to acknowledge their coaches or PLC members who shaped solutions within the context of the particular classroom. The following is a quote of Interviewee #1;

- My coaching teacher at this school utilized many techniques from my studies as part of her classroom practice such as NCTM's 5 Practices (2011). It very much benefited me that my educational program fit hand in glove with the philosophies of my in-class mentor.

Interviewee #2 stated:

- I feel that here, the school level PD needs to be integrated into the grade level and subject specific PLC groups because what we really need to do is to go from theory to practice.

Interviewee #5 stated:

- Coaching is so, so important. Administration and coaches need to be a safe space for teachers, and my evaluator needs to be able to say what is going to happen with a growth mindset.

Interviewee #6 stated:

> - *I have learned from the Professional Development at the school, but sometimes the Professional Development times are not so useful since they are focused on other subject areas.*

Teachers did not talk specifically about trust-based social supports that create group efficacy (Donohoo, Hattie, & Eells, 2018), but as related in the quote from Interviewee #2 above, teachers do mention working well with team members as being a source of a positive and productive work environment.

Conclusion

In October 2019, the state DoE approved the school's request to remain open to serve its community and continue in its path as an Innovation Status School. In this decision, the DoE acknowledges the positive changes resulting from the newly implemented innovations and structures. The state noted that overall, the school continues to struggle in some critical areas, but has improved markedly in a number of ways. Those improvements include a 50% reduction in student expulsions and a decrease in staff turnover from over 50% to a little over 20% in the last couple of years (CDE, 2019). During site visits, the state noted an improved focus on instructional rigor. They also noted how primary academic indicators of Mean Growth Percentiles have moved upwards from their historic lows, although still low in all areas.

The DoE report also mentions the large size and historical role of the school as a community anchor in its decision to forestall closure of the school (CDE, 2019). The report noted that, "With 2,000 students enrolled, there is not enough enrollment capacity to enroll these students at other schools" (CDE, 2019, p. 4). The report notes that the school serves a unique population of students including a significant number of ELLs, newcomers and refugees, students experiencing homelessness, and transient students (e.g., more than 400 new students enrolled mid-year during the last school year), and that the school "provides unique programming to support these students" (CDE, 2019, p. 3). Overall, the recommendations of the state are to give the process a little more time and to stay the course.

Nevertheless, the state has persisted in its demand for further academic improvement and leaves the ultimate fate of the school to a future decision. For the mathematics department, future improvements depend upon addressing weak points that have previously eluded the turnaround focus. One such phenomenon revealed in the data is the disparity between turnover rates of the school as a whole and turnover rates within the mathematics department. While regular classroom teacher turnover rates have improved to about 20% overall, the mathematics department continues to experience about 50% yearly turnover in teachers. Since teacher experience and preparation is highly related to student success (United States Department of Education, n.d.), it is worth considering whether the unique forces acting upon the mathematics department could be mitigated.

Suggestions gleaned from research point to how in turnaround schools with a narrow focus on test scores diminish teacher autonomy over the technical aspect of their work (Gehrke, 2005; Hughes, 2006; Santoro, 2018). When experienced teachers are not allowed to make curricular or instructional decisions, their self-efficacy decreases since in their perception, they are not able to respond with their own creative expertise to meet the needs of students. Further, for the veteran teacher, the many accountability systems that are in place in a turnaround school can seem punitive rather than supportive (Santoro, 2011).

On the other hand, mathematics teachers with little experience and less formal background in educational pedagogy might see the daily oversight and narrow instructional focus as a boon since they seek structure and guidance in a way that is met by daily oversight and lack of autonomy. Also, less experienced teachers may not be highly impacted by rigid accountability systems since their personal history of past evaluations is small or even non-existent. A more thorough understanding of these issues within the mathematics department holds potential to allay teacher stress, increase teacher self-efficacy, and ultimately, increase achievement for language learners and for all students of mathematics.

References

Colorado Department of Higher Education College Opportunity Fund (COF) stipend page. (n.d.). Retrieved December, 2019 from https://highered.colorado.gov/Finance/COF/.

Colorado DoE. (2019, October). State review panel progress monitoring recommendation form (by Schoolworks, Inc.).

Colorado DoE Website. (2019). *Schoolview*. Retrieved December 20, 2019, from https://edx.cde.state.co.us/SchoolView/DataCenter/reports.jspx?_adf_ctrl-state=pac20phbp_4&_afrLoop=9158046608723947&_afrWindowMode=0&_adf.ctrl-state=1bj88fdc94_4.

Data USA. Mathematics teacher education page (n.d.). Retrieved December, 2019 from https://datausa.io/profile/cip/mathematics-teacher-education.

De La Garza, G. (2016). *Aurora central high school, Aurora public school district innovation application to the Colorado DoE*. Retrieved from http://aps2020-archive.aurorak12.org/wp-content/uploads/sites/176/2016/02/Central-Innovation-Plan-Final-Draft-Feb-29-2016.pdf.

Donohoo, J., Hattie, J., & Eells, R. (2018). Leading the energized school. *Educational Leadership, 75*(6), 40–44. http://www.ascd.org/publications/educational-leadership/mar18/vol75/num06/The-Power-of-Collective-Efficacy.aspx.

Fantuzzo, J., Perlman, S., Sproul, F., Minney, A., Perry, M., & Li, F. (2012). Making visible teacher reports of their teaching experiences: The early childhood teacher experiences scale. *Psychology in the Schools, 49*(2), 194–205.

Gehrke, R. S. (2005). Poor schools, poor students, successful teachers. *Kappa Delta Pi Record* (Fall 2005), 14–17. https://files.eric.ed.gov/fulltext/EJ724902.pdf.

Guglielmi, R., & Tetrow, K. (1998). Occupational stress, burnout, and health in teachers: A methodological and theoretical analysis. *Review of Educational Research, 68*, 61–99.

Hansen, M., & Choi, K. (2011). *Chronically low-performing schools and turnaround: Findings in three state.* National center for analysis of longitudinal data in education research. Washington, DC: Calder Center. Available at http://www.caldercenter.org/sites/default/files/wp-60.pdf.

Herman, K. C., Hickmon-Rosa, J., Reinke, W. M. (2018). Empirically derived profiles of teacher stress, burnout, self-efficacy, and coping and associated student outcomes. *Journal of Positive Behavior Interventions, 20*(2), 90–100. Available at https://doi.org/10.1177/1098300717732066.

Hughes, J. C. (2006). *Teacher stress, teacher efficacy, and standarized testing: A study of New York City public school teachers*. Dissertation abstracts international-A, 67(03), AAT 3210270.

Kristensen, T. (1995). The demand-control-support model: Methodological challenges for future research. *Stress Medicine, 11*(1), 17–26. https://doi.org/10.1002/smi.2460110104. Retrieved at https://www.researchgate.net/publication/229876354_The_Demand-Control-Support_Model_Methodological_Challenges_for_Future_Research.

National Council of Teachers of Mathematics Position Paper. (2014). *Access and equity in mathematics education*. Retrieved from https://www.nctm.org/uploadedFiles/Standards_and_Positions/Position_Statements/Access_and_Equity.pdf.

Ouellette, R. R., Frazier, S. L., Shernoff, E. S., Cappella, E., Mehta, T. G., Marinez-Lora, A., & Atkins, M. S. (2018). Teacher job stress and satisfaction in urban schools: Disentangling individual, classroom, and organizational level influences. *Behavior Therapy, 49*(4). Retrieved December 2019, from https://www.researchgate.net/publication/321861656_Teacher_Job_Stress_and_Satisfaction_in_Urban_Schools_Disentangling_Individual_Classroom_and_Organizational_Level_Influences.

Payne, R., & Fletcher, B. C. (1983). Job demands, supports, and constraints as predictors of psychological strain among school teachers. *Journal of Vocational Behavior, 22*, 136–147. Retrieved from https://fdocuments.in/document/job-demands-supports-and-constraints-as-predictors-of-psychological-strain.html.

Robles, Y. (2018, August 13). El distrito de Aurora se prepara para ofrecer más servicios de traducción e interpretación. *Chalkbeat, Chalkbeat en español*. Available at http://go.chalkbeat.org/webmail/342281/163811823/1326822b796ea9de55b84f16c3353ab9f31c6fae70ed908fa6fd1666b75f546c.

Royster, P., Gross, J., & Hochbein, C. (2015). Timing is everything: Getting students back on track to college readiness in high school. *The High School Journal, 98*(3), 208–225. Project MUSE. https://doi.org/10.1353/hsj.2015.0005.

Santoro, D. A. (2011). Good teaching in difficult times: Demoralization in the pursuit of good work. *American Journal of Education, 118*(1), 1–23.

The University of Chicago Press. Retrieved from. https://www.jstor.org/stable/10.1086/662010.

Santoro, D. A. (2018). *Demoralized: Why teachers leave the profession they love and how they can stay*. Cambridge, MA: Harvard Education Press.

Smarick, A. (2010). The turnaround fallacy. *Education Next, 10*(1). Retrieved from https://www.educationnext.org/the-turnaround-fallacy/.

Stuit, D. (2010). *Are bad schools immortal? The scarcity of turnarounds and shutdowns in both charter and district sectors*. Washington, DC: Thomas B. Fordham Institute. Retrieved from https://fordhaminstitute.org/sites/default/files/publication/pdfs/fordhamimmortal10.pdf.

Tsoi-A-Fatt Bryant, T. (2015). *College preparation for African American students: Gaps in the high school educational experience*. Washington, DC: CLASP. Retrieved from https://www.clasp.org/sites/default/files/public/resources-and-publications/publication-1/College-readiness2-2.pdf.

United States Department of Education. (n.d.). Improving teacher preparation: Building on innovation. Retrieved December, 2019 from https://www.ed.gov/teacherprep.

Yang, X., Ge, C., Hu, B., Chi, T., & Wang, L. (2009). Relationship between quality of life and occupational stress among teachers. *Public Health, 123*(11), 750–755.

Yu, X., Wang, P., Zhai, X., Dai, H., & Yang, Q. (2015). The effect of work stress on job burnout among teachers: The mediating role of self-efficacy. *Social Indicators Research, 122*(3), 701–708. https://doi.org/10.1007/s11205-014-0716-5.

12

Feedback and Assessment for Emerging Bilinguals in Secondary Mathematics: A Case Study of Written Feedback

Rachel S. G. Bower and Vanessa Z. Mari

Teachers put a lot of effort not only into creating assessment opportunities for their students, but in providing feedback for the work that they do. Getting students to engage with feedback has long been a desire for mathematics teachers. This looks different if the students are learners of English. Students, if they even attempt to access teacher feedback, must navigate mathematical and non-mathematical notations and abbreviations. Emergent bilingual (EB) students who attempt this navigation have the added complexity of doing so in a second language. Through this case study we analyzed the type of written feedback that emergent bilinguals students receive in their mathematics assessments in order to document best practices. For the purpose of this research, we define teacher generated written feedback as "any comment, question, or

R. S. G. Bower (✉)
Las Vegas, NV, USA
e-mail: rachel.bower@nsc.edu

V. Z. Mari
Nevada State College, Henderson, NV, USA
e-mail: vanessa.mari@nsc.edu

L. C. de Oliveira and M. Civil (eds.), *Teaching Mathematics to English Language Learners*, https://doi.org/10.1007/978-3-030-48355-5_12

error correction written on students' assignment" (Mack, 2009, p. 33). We wondered what research shows could be done to improve assessment and teacher feedback for EB students in the secondary mathematics classroom. We also wondered how close or wide of the mark current classrooms are to these research recommendations.

Researching Assessment

When we explore the type of written feedback that emergent bilinguals receive on their different mathematical assessments, it is important to also consider what research says about the assessments themselves. Research in assessment has focused on its multiple strands, which include formative and summative assessments that diverge into different areas such as oral, written, electronic, performative assessment, among many others. Lee and Son (2015) define assessment literacy as teachers' ability to "identify, select, or create assessments optimally designed for various purposes, such as accountability, instructional program evaluation, student growth monitoring and/or promotion, and diagnosis of specific student needs" (p. 48). As it is described in the definition, we see a shift from what assessments means in education toward a focus on what teachers can do with the data they generate is now more prominent.

Historically and societally, the purpose of assessment is to assign grades and label students as learning or not learning. For educators, the role of assessment has broadened as has the fodder. Lee and Son's definition of assessment literacy demonstrates the varied uses of assessment in the classroom. We would add to this list an additional purpose of assessment is to allow educators to gauge the effectiveness of their lesson planning ad make adjustments accordingly. The National Council of Teachers of Mathematics (NCTM) process standards (NCTM, 2000) and the Common Core State Standards for Mathematics (CCSSM) eight mathematical practices (CCSSM, 2010) encourage the use of various assessment types like oral or project based. So now, instead of featuring a list of similar, abstract problems to solve, assessments might offer students the opportunity to collaborate with a peer, justify their work in writing, or represent their work in multiply ways. These changes in assessment can afford students, even in mathematics, the chance to demonstrate their language and writing skills. These broader possibilities for language in assessments can be seen in Takanishi and Le Menestrel's (2017) *Promoting the Educational Success of Children and Youth Learning English: Promising Futures.* In their book they expand on what they term, "Promising Practices," when working with secondary EBs (pp. 317–325). These nine practices are listed below.

1. Develop Academic English as Part of Subject-Matter Learning
2. Integrate Oral and Written Language Instruction into Content Area Teaching
3. Provide Regular Structured Opportunities to Develop Written Language Skills
4. Develop Reading and Writing Abilities of ELs Through Text-Based, Analytical Instruction Using a Cognitive Strategies Approach
5. Provide Direct and Explicit Comprehension Strategy Instruction
6. Provide Opportunities for Extended Discussion of Text Meaning and Interpretation
7. Foster Student Motivation for and Engagement in Literacy Learning
8. Provide Regular Peer-Assisted Learning Opportunities
9. Provide Small-Group Instructional Support for Students Struggling with Literacy and English Language Development.

In 2015, a National Science Foundation sponsored research project examined effective teaching practices for EB students in middle school mathematics classrooms. The research showed that effective teachers connect mathematics with students' life experiences, connect mathematics with language, use visuals, and discuss students' mathematical writing (Sorto, Wilson, & White, 2018). These recommendations reinforce the emphasis on English literacy we see from Takanishi and Le Menestrel (2017) and should be incorporated into high-quality assessments.

NCTM, similar to Lee and Son (2015), states that assessment should serve four functions: monitoring students' progress, making instructional decisions, evaluating students' achievement, and evaluating programs

(2014, p. 89). Current research encourages the mathematics education community to de-emphasize assessment as a vehicle for assigning grading and increase formative assessment experiences that arise from engagement with substantive, mathematical tasks (Boaler, 2016; NCTM, 2014). In addition, the students should be expected to regularly assess themselves against the learning goals for a unit.

This brief foray into research recommendations and best practices shows there are many considerations that mathematics teachers need to take into account when developing assessments. Not only do they need to be developed to assess student knowledge, but also as a way to expand their own practices and further assist the learners. Quality assessments should provide opportunities for students to connect mathematics with real-world contexts, utilize visuals, and allow for writing about the mathematics. When it comes to serving the EB students in the classroom, we were also interested in seeing the ways that teachers provide written feedback to their EB students.

Researching Feedback

NCTM recommendations highlight the importance of how students experience mathematics because this has a significant impact on the ways in which they identify themselves as doers of mathematics (2018, p. 5). One of the ways that students experience mathematics would be the type of written feedback they receive in school. This written, teacher generated feedback influences how they identify themselves as doers of mathematics since it can be perceived as encouraging or discouraging. This is one of the reasons why exploring the feedback that students receive is of importance. Researchers of EB students and mathematics have made recommendations such as giving students, "feedback that purposefully focuses on mathematical language development, and allows students to revise their writing" (Ramirez & Celedón-Pattichis, 2012, p. 23). A focus on mathematical language use in written feedback is important because it provides students with more opportunities to practice and correct their academic language use. Finally, research in this area has also focused on the value of diagnostic comments as a form of feedback. It was found that students who received diagnostic comments and no grades achieved at a significantly higher level in mathematics (Boaler, p. 143). This highlights the importance that written feedback has on student achievement and the correlation that this has with how they perform.

Francis and Stephens' (2019) recently published *English Learners in STEM Subjects: Transforming Classrooms, Schools, and Lives* which includes the following recommendations in its chapter titled, Assessing STEM Learning among English Learners:

- Utilize multiple types of assessment, including oral discourse
- Ensure rubrics are understandable and culturally sensitive
- Make use of multiple representations and scaffolding, such as visual supports and sentence frames
- Teachers need to interact with and know their students
- Encourage self/peer-assessment and foster self-directed learners
- Feedback should be effective for learning, i.e., more than praise or a red 'X'
- Use assessment results to make changes to classroom instruction.

Here we see similar sentiments as our previous sources concerning the nature of assessment we want to use in mathematics classrooms and the purpose of the assessment. We also have additional encouragement to provide diagnostic feedback and allow student opportunities for self-assessment. Thorough teacher feedback and student self-reflection classrooms can encourage students to engage with mathematical ideas and writing. All of these serve to promote the growth mind-set culture that is championed by Carol Dweck, Jo Boaler, and others (Growth Mindset, n.d.) and encourage the inclusion of literacy learning in mathematics classrooms.

Assessment at the classroom level might be the best place to evaluate EB mathematical understanding as large-scale, popular benchmark assessments may not be valid for EB students (Francis & Stephens, 2019, p. 10). Therefore, to explore assessment and feedback we partnered with two secondary mathematics teachers from similar large, urban high schools. Each teacher has a significant population of EB students. We then pieced together case studies to analyze the language being used in written feedback and the nature of assessments. We felt it was important to do this as,

> much of the research on classroom summative and formative assessment has been conducted without considering linguistic diversity in the classroom. With a few exceptions, research is mostly silent about the linguistic factors that may shape effectiveness in formative assessment, or it appears to implicitly assume equal proficiency in the language of instruction among all students in class. As a consequence, the extent to which current knowledge of formative assessment is applicable to linguistically diverse classrooms still needs more research. (Francis & Stephens, 2019, p. 209)

It is our hope that our case study adds to and continues the conversation of best classroom practices for EBs as we continue to explore the role of language in mathematic assessments and feedback.

A Tale of Two Teachers

"Couldn't Ask for Better Students."

Anthony is a first-year teacher at a large, urban high school where 90% of the students come from historically underrepresented populations and 73% are classified as Latinx. Anthony earned a bachelor's degree in philosophy, but quickly decided to pursue a teaching license as a postgraduate. He was hired on a provisional license to teach high school mathematics. Like many first-year teachers, he was given the lowest level mathematics classes in high school which include Pre-Algebra and Algebra. Two of his five sections are comprised entirely of EB students that are a mix of Spanish and Swahili speakers. Anthony knows Portuguese and is in the process of learning Spanish.

When he was hired and learned he would be working with EB students, Anthony began brushing up on his math and Spanish using Duolingo and watching Khan academy videos, in Spanish. He even moved into an apartment across the street from the school to fully immerse himself in the campus culture. When we met Anthony, it was at the end of his first-year teaching and his experiences echo those of first-year teachers everywhere. He works, a lot. He hasn't taken the time to collaborate with colleagues and he believes this would have been beneficial. He feels the professional development he has received related to EB students emphasizes other content areas and age groups.

Anthony has misgivings about choices he has made in the classroom. In the future he will advocate more for EB students who he believes have been placed in the wrong math class. He believes speaking too much Spanish has provided a crutch for EB students. "I thought it was someone else's job to teach them English." To that end, Anthony purposefully used very little English words, that is words of any kind, in his assessments.

"My Love Is Algebra."

As part of our case study, we also worked closely with Callie. She is a multigenerational math teacher; her grandmother taught in a one-room schoolhouse. She teaches at a large, urban school a mere 4 miles from Anthony. Her school is also composed of 90% students from historically underrepresented groups with 67% of them being Latinx. Callie is a highly educated, veteran teacher with 19 years of service. Callie currently teaches sections of Algebra I and College Prep Math with EB students learning beside native English speakers. The vast majority of her EB students have been Spanish speakers.

Callie is a confident teacher, describes herself as, "strict," and is likely to rail against policies from administrators or the state that she does not believe advance education in her classroom. Callie has accumulated lots of experience over the years in teaching EBs, but interestingly does not acknowledge her Teacher of English as a Second Language (TESL) endorsement because she does not want to be, "pigeonholed as a teacher of EB students." She continues on to say, "as soon as you have that you have to teach that class because no one else has that endorsement."

Callie is aware of and uses many accommodations for EB students such as bilingual dictionaries, cognates, manipulatives, peer tutors, and encourages native language use with the curriculum. She modifies assessments mainly by using assessment menus and allows for retakes. Although she believes word problems are a staple in her class work, notes, and homework they do not feature prominently in her assessments.

Analysis

Anthony and Callie shared with us different assessments they had given their students. Anthony provided us with copies of all quizzes, tests, and exams he administered and graded for all of his classes for a marking period. Callie did the same, but was able to share more of her assessments as she saved them since the start of the school year. These two teachers do not work at the same school and did not use the same assessments, although both these teachers work in schools with very similar student demographics and large percentages of Latinx students. Although these teachers are at significantly different places in their careers, there is not a noticeable difference in their assessments or the written feedback they provide.

We anticipated employing a painstaking organizational matrix in order to make sense of the written feedback from Anthony and Callie's assessments, but that proved impossible. Even with the large collection of samples they provided, the total written feedback was sparse. Anthony's assessments are graded, mistakes feature a traditional 'X' in dark ink. Any written feedback almost entirely references a missing or incomplete "check" of the solution as shown in Fig. 12.1. In Fig. 12.2, we see commentary concerns the reasonableness of work done while checking a solution, but a signage issue may confuse readers.

As previously stated, Anthony purposefully wrote assessment items without context to keep his students from struggle with language issues. Although Callie claimed to employ contextualized problems with her students during class time and in homework assignments these did not translate to her assessments. The one word problem we were able to see from Callie, Fig. 12.3, was only attempted by about half the students

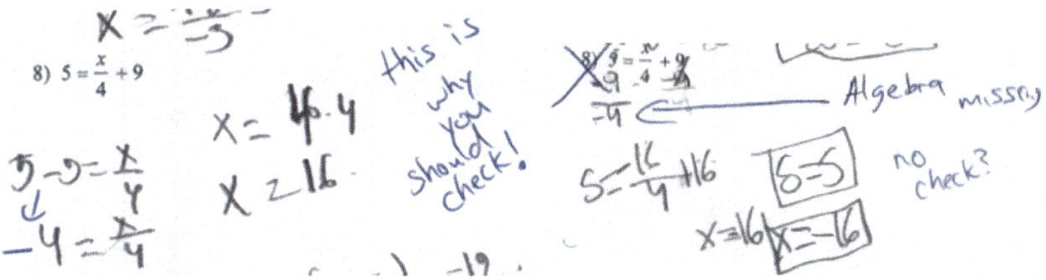

Fig. 12.1 Written feedback encourage checking the solution

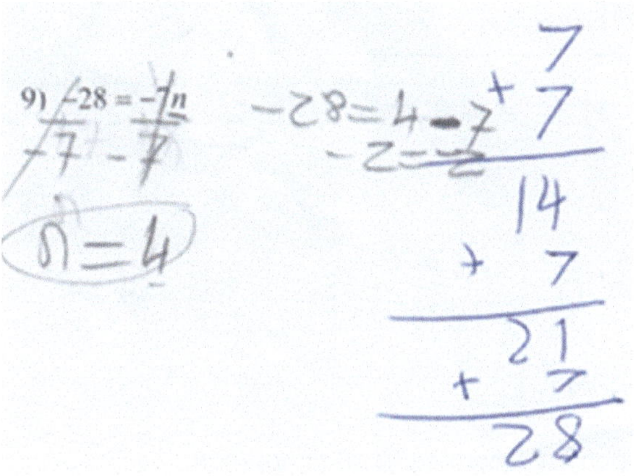

Fig. 12.2 Written feedback reasonableness of an answer

8) The state fair is a popular field trip destination. This year the senior class at High School A and the senior class at High School B both planned trips there. The senior class at High School A rented and filled 4 vans and 1 bus with 89 students. High School B rented and filled 9 vans and 3 buses with 225 students. Each van and each bus carried the same number of students. How many students can a van carry? How many students can a bus carry?

Fig. 12.3 Word problem example

and solved correctly by half of attempters. Callie is almost as frugal as Anthony with written feedback, often assessments bear only a score, circled item, or a question mark as in Fig. 12.4. The feedback she does leave is often used to stress proper mathematical notation as in Fig. 12.5. In Fig. 12.6, we see examples of the other use Callie makes of written feedback which is to communicate with her students.

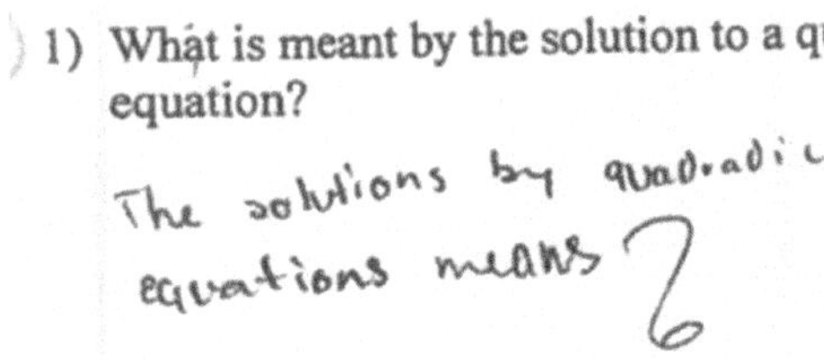

Fig. 12.4 Written feedback limited to a question mark

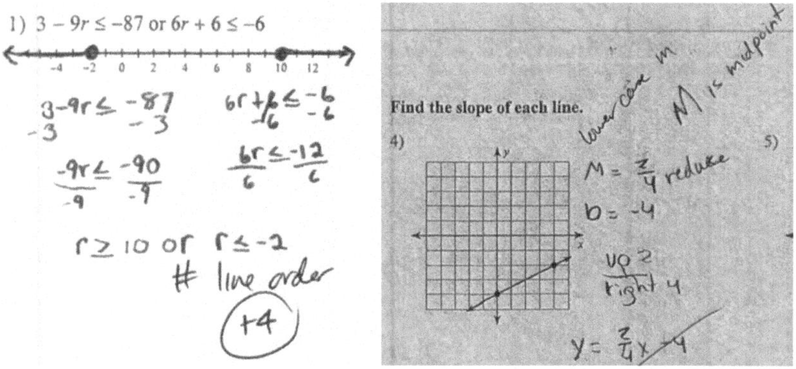

Fig. 12.5 Written feedback emphasizing notation

Language Recommendations for Anthony and Callie

Both teachers admit contextualized or word problems are rare in their assessments, but contextualized problems offer an opportunity to engage with language and are more likely to allow for multiple methods of solving and written responses. Engaging with authentic mathematical tasks can be more meaningful than solving pages of abstract calculations (Boaler, 2016). A silver lining of contextualized problems is that they are harder to solve using an app or website. The use of these kinds of math apps was a specific complaint we heard from Callie. Moreover, research tells us that contextualized problems are part of effective instruction for EB students. The one true contextualized problem we saw, Fig. 12.3, could be written in more concise language and allow

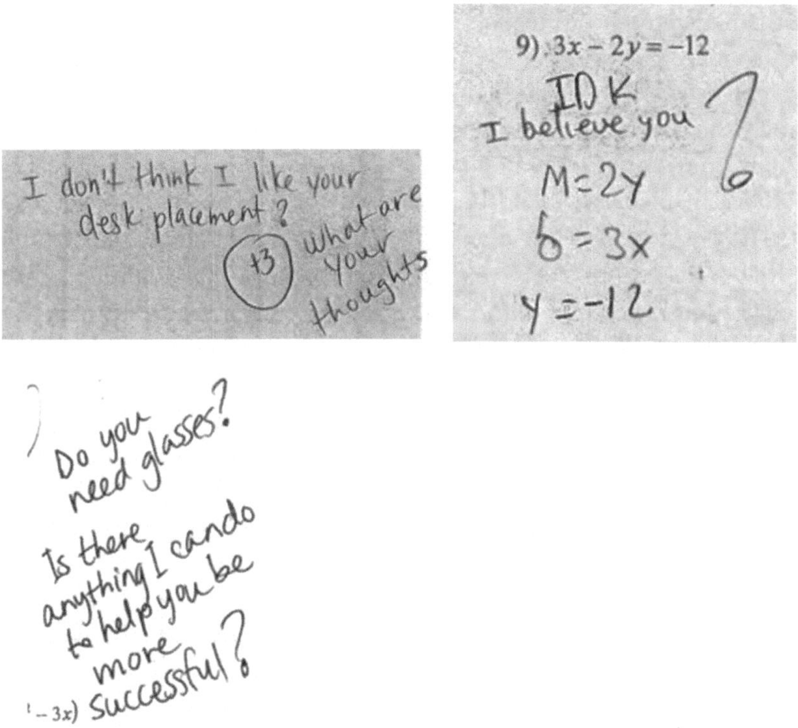

Fig. 12.6 Communicating with her students

enough space to encourage various problem-solving strategies. It could also include graphics to help decipher certain words. Sentence stems are a useful tool for EB students when considering contextualized problems. A sentence stem for Fig. 12.3 might be, "____ students can ride in a bus. _____ students can ride in a van." Similarly, a sentence stem could improve performance in Fig. 12.2 which shows an EB student attempt to construct a stem, but never engage with the mathematics. A sentence stem here might say, "The solution to a quadratic equation ___________." Including an *XY*-plane and a generic parabola with Fig. 12.4 would further check for understanding by allowing students to mark the solutions. Students might know the solutions are where the graph intersects the *x*-axis, but not yet be able to express this in words.

Mathematics teachers are evaluating both mathematical understanding and language gains.

Some teachers may believe that providing substantive written feedback is too time consuming a proposition, but we found these teachers spent a lot of time writing the same terse feedback on *every* paper. Figure 12.1 and similar questions could be organized so that the check has a dedicated space and is ready to receive the variable's value for example, CHECK: $5 = \frac{()}{4} + 9$. This reminds the students that completing the check is the expectation without having to scrawl it on 100 + papers. A dedicated space for a required check keeps the mathematics a bit neater so that the solution and check both have room. It also helps the teacher determine if students know what the check step requires, are having trouble with the arithmetic, or maybe are choosing not to do it. It is harder to forget if the reminder is preprinted on the paper! The same can be said for mathematical notation such as we see in Fig. 12.5. If students are asked to find slope, why not add, "slope (m) = ———?" A vinculum can remind students the answer should be left as a fraction if that is a requirement. Some educators may say we are indulging students when we do these things, but adding such aids to our assessment items can help teachers understand where students are having difficulties and decrease language barriers. If students consistently see assessment items presented this way, they might start to write them in when absent. Furthermore, such aids might assist students in end of course standardized testing where multiple choice answers and distractors have similar verbiage.

When teachers begin to leave meaningful feedback on assessments the emphasis on slashes, the red 'X,' question marks, checks, and percentages correct will be decreased. This is an important step toward creating a supportive learning community where understanding is the goal and mistakes are viewed as learning opportunities. We see Callie demonstrating support of her students' learning through written feedback in Fig. 12.6. Teachers who sometimes feel they don't always have the time for descriptive, written feedback might instead opt to highlight mistakes. Then allow class time for students to work together to revise their highlighted work and resubmit. These kinds of opportunities with peers foster student mathematical discourse and discussion of student written work. Additionally, teachers can create one document where they address all of the most common errors students had. This way the students have a written record they can reference in the future.

General Language Recommendations

When it comes to recommendations for classroom practice, it is not enough to simply cite the research. Practitioners deserve insights on how to implement such changes, sometimes significant ones, with their own students. We often hear the same kinds of questions from our students, attendees at mathematics education and Teaching English to Speakers of Other Languages (TESOL) conferences as well as locally led professional development. Questions like, *How can I provide more written feedback given the time constraints in a secondary classroom?* or *How do I incorporate writing and language? I teach math!* We provide the following as a way to address the common questions we hear about teaching mathematics with EB students. We encourage e-mails from readers with additional questions or comments concerning their experiences on the topics of assessment and feedback. The most important asset we have as educators is each other.

Educators often will discuss their impressions after a whole class assessment, maybe it feels expedient to provide feedback orally. Hey, we still do this ourselves! However, EB students require multiple forms of language input in order to make connections between the task and the language. When the assessment is discussed orally but also in writing, there are multiple ways for the students to make these connections and reinforcement with the language. Additionally, if the student would like to revise the work at a later time, they have annotations and feedback from the teacher than can help them in this process. Providing written feedback is also a practice that is useful for native English-speaking students.

We have already discussed why it is important to provide written feedback. Giving oral feedback is great, but we also need to focus on providing support in all language skills. We recognize that writing feedback on every paper can be a huge time commitment, but there are ways to provide more written support for students. One recommendation is to find out which of the assessment items posed the greatest challenge. Once the teacher recognizes which are these items, they can create a new document with feedback on how to complete those specific items. That way, all students receive further written reinforcement on the tasks that gave them more difficulty. Another recommendation would be to engage in peer feedback. Students can be placed with partners that share similar linguistic abilities and they can provide each other feedback with the teacher's assistance. This way they are also using their oral academic vocabulary to discuss work from class.

One of the things we noticed through these case studies was a limited variety in assessment items students are presented. They did not require the students to use written language, contained very little text and mostly focused on solving equations procedurally. It is important to develop test items that use pictures, language, diagrams, and other mathematical representations for students to see as support. For example, teachers can include a partially completed Frayer model. A Frayer model is a type of graphic organizer that helps students build on vocabulary concepts by having them define, create examples, and describe characteristics of a concept. A simple internet search yields numerous examples and templates. Teachers can use a version of this with assessment items and communicate the expectation that students finish the model. The expectations of the Frayer model may vary depending on the problem asked, but can include titles such as: model, sketch, justify your answer in words, define, equation, table, or graph. Such short phrases can also be easily translated for students needing that support.

We noticed that disproportionately, many EB students did not complete word problems or even attempted to work on them. Reasons for this likely include not understanding the language and the concepts being asked. Providing language support in assessments for EB students might sound like an onerous task, but it doesn't have to be. When creating assessments think about what have been the new concepts discussed during the unit and what new language has been used. We recommend that teachers provide with their assessments a list of vocabulary words with their definitions. This list can appear on the first page of the assessment and students can use this to refer back as they work. Another recommendation is assisting students with building their own mathematics bilingual dictionaries which they can refer to when working on their assessments. Bilingual dictionaries are a great way to support students' learning of English as well as learning academic vocabulary in their native language. Working with peers can be a great language support even when the assessment is administered individually. Students should be allowed sufficient time to work together during class time on tasks similar to what they will see on assessments. Even after an assessment when they are reflecting and making corrections, this can be done collaboratively.

The longer I teach mathematics the further removed I am from that English 101 course I took as an undergraduate. I never felt as comfortable in writing-intensive courses as I did in my mathematics courses and a lot of math educators might also feel this way. Thus the idea of incorporating writing and literacy in their math teaching can feel daunting to a lot of educators, but it doesn't have to. Many effective instruction strategies allow for students to exercise their literacy muscles in the pursuit of mathematics learning and are encouraged by current standards. For example, asking students to self-reflect about what they are learning is a great opportunity for them to write about mathematics. We can also create assessment items that encourage more written responses such as presenting students with common mistakes and ask them to, in writing, explain to a friend how to fix the error. Communicating and collaborating are par for the course in the work of mathematicians, but should also be the expectation for young mathematicians.

Our recommendations encourage students working together, having their language needs exercised and supported, and receiving substantive feedback. We want students to feel rewarded for their efforts. Through this research we noticed how the students that received most of the reinforcements or positive affirmations were students that were already performing well. These were the students that received a smiley face, a heart, or a positive affirmation. We think this is great practice because it continues to motivate the students that are performing well. On the other hand, we argue that these affirmations should also be present with work that is not complete. Recall Callie, a teacher in this case study, wrote on the assessment of one student, "How can I help you succeed?"

This is a great example of providing positive reinforcement to students and fostering a supportive learning environment.

Conclusion

When we are working with EB students, we need to take into consideration that they need input in all of their language skills. Providing oral feedback is an excellent way to have a conversation with students regarding their decision making when doing mathematics. On the other hand, we argue that oral feedback needs to be accompanied with quality assessment and substantive, written feedback. Having access to the teachers' written feedback will help students self-evaluate their work even when they are not at school. During this chapter we discussed the different ways that written feedback can be modified for EBs as well as ways to make this easier for the teachers to accomplish.

We feel there was and is value in examining the assessments and feedback we are using with all mathematics students, particularly EB students. One way teachers can do this is to seek out professional development related to EB students and their needs (Pettit, 2011). Professional development can happen anytime teachers come together to form professional learning communities (PLCs) on their campuses and take the opportunity to learn from each other to create quality assessments and feedback.

References

Boaler, J. (2016). *Mathematical mindsets: Unleashing students' potential through creative math, inspiring messages, and innovative teaching*. San Francisco, CA: Jossey-Bass.

Executive Summary. (2018). *Catalyzing change in high school mathematics: Initiating critical conversation.* Reston, VA: NCTM.

Francis, D., & Stephens, A. (Eds.). (2019). *English learners in STEM subjects: Transforming classrooms, schools, and lives.* Washington, DC: The National Academies Press.

Growth Mindset. (n.d.). Retrieved from https://www.youcubed.org/resource/growth-mindset/.
Lee, J., & Son, J. (2015). Two teacher educators' approaches to developing preservice elementary teachers' mathematics assessment literacy: Intentions, outcomes, and new learning. *Teaching and Learning Inquiry: The ISSOTL Journal, 3*(1), 47–62. Indiana University Press.
Mack, L. (2009). Issues and dilemmas: What conditions are necessary for effective teacher written feedback for ESL learners? *Polyglossia, 16,* 33–39.
National Council of Teachers of Mathematics. (2000). *Principles and standards for school mathematics.* Reston, VA: NCTM.
National Council of Teachers of Mathematics. (2014). *Principles to action: Ensuring mathematical success for all.* Reston, VA: NCTM.
National Governors Association Center for Best Practices, & Council of Chief State School Officers. (2010). *Common Core State Standards Mathematics.* Washington, DC: National Governors Association Center for Best Practices, Council of Chief State School Officers.
Pettit, S. (2011). Teachers' beliefs about English language learners in the mainstream classroom: A review of the literature. *International Multilingual Research Journal, 5*(2), 123–147.
Ramirez, N. G., & Celedón-Pattichis, S. (2012). Second language development and implications for the mathematics classroom. In S. Celedón-Pattichis & N. G. Ramirez (Eds.), *Beyond good teaching: Advancing mathematics education for ELLs* (pp. 19–37). Reston, VA: NCTM.
Sorto M. A., Wilson A. T., & White A. (2018). Teacher knowledge and teaching practices in linguistically diverse classrooms. In J. Moschkovich, D. Wagner, A. Bose, J. Rodrigues Mendes, & M. Schütte (Eds.), *Language and communication in mathematics education.* ICME-13 Monographs. Cham: Springer.
Takanishi, R., & Le Menestrel, S. (Eds.). (2017). *Promoting the educational success of children and youth learning English: Promising futures.* Washington, DC: The National Academies Press.

Index

L. C. de Oliveira and M. Civil (eds.), *Teaching Mathematics to English Language Learners*, https://doi.org/10.1007/978-3-030-48355-5

The manufacturer's authorised representative in the EU is Springer Nature Customer Service Centre GmbH, Europaplatz 3, 69115 Heidelberg, Germany. If you have any concerns regarding our products, please contact ProductSafety@springernature.com

Printed and bound by CPI Group (UK) Ltd, Croydon, CR0 4YY

30/09/2026

02225646-0001